Marine Electrical Technology

Marine Electrical Technology

Gideon Paxton

RANDOM PUBLICATIONS
NEW DELHI (INDIA)

Marine Electrical Technology

ISBN 978-93-5111-939-5

Published in 2016 in India by

RANDOM PUBLICATIONS

4376-A/4B, Gali Murari Lal, Ansari Road
New Delhi-110 002
Phone : +9111-43580356, 011-23289044, 011-43142548
e-mail: sales@randompublications.com,
info@randompublications.com, randomexports@gmail.com

Reprinted 2025

Type Setting by : Friends Media, Delhi-110089
Digitally Printed at : Replika Press Pvt. Ltd.

Preface

The field of automation and control system is vast; there are many different ways of achieving the required effects by the operator of marine systems. It is impossible to cover all the aspects of the control used in ship.

Electricity has to be distributed with minimal losses after generation. Unlike a shore based transmission system, where the lengths of the conductors run throughout a country, a shipboard electrical distribution system is short and simple. In this book, the general layout of the main electrical distribution system, along with the main switchboard and emergency switchboard arrangements.

The electrical requirements of the training ship are supplied by the electrical plant. The electrical plant consists of three turbo-generators, one emergency generator, the emergency battery system, and the electrical distribution system. There are three 750 kW turbo-generators, arranged for individual and parallel operation with any two capable of carrying the maximum sea and port load of the ship and the third generator available for standby. Each turbo-generator is capable of a continuous rating of 750 kW, with a 25% overload for two hours. A 200 kW emergency diesel generator is adequate for supplying power for emergency lighting, interior communication and power load requirements and also cold ship starting of the plant.

Paralleling of main and emergency generators is prevented by an interlocking circuit which trips the bus tie breaker on the main switchboard and emergency switchboard when the generator breaker is closed. One emergency battery bank, rated at 120 volts, 204 ampere-hours is provided for the ships emergency lighting, 120 V DC bus, fire door release system, diesel starting control circuit and the manual alarm and fire detection system.

The book presents current development in a variety of areas that impact offshore and marine engineering. This book is very useful not only for marine engineering students but also for research scholars and teachers.

– Author

Contents

1

Ship's Electrical System

A ship is like a floating city with all the privileges enjoyed by any normal land city. Just like a conventional city, the ship also requires all the basic amenities to sustain life on board; the chief among them is power or electricity. In this article we will learn as to how power is generated and supplied on board a ship.

POWER GENERATION ON BOARD

Shipboard power is generated using a prime mover and an alternator working together. For this an alternating current generator is used on board. The generator works on the principle that when a magnetic field around a conductor varies, a current is induced in the conductor.

The generator consists of a stationary set of conductors wound in coils on an iron core. This is known as the stator. A rotating magnet called the rotor turns inside this stator producing magnetic field. This field cuts across the conductor, generating an induced EMF or electro-magnetic force as the mechanical input causes the rotor to turn. The magnetic field is generated by induction (in a brushless alternator) and by a rotor winding energized by DC current through slip rings and brushes.

Few points to be noted about power on board are:

- AC, 3 phase power is preferred over DC as it gives more power for the same size.
- 3 phases is preferred over single phase as it draws more power and in the event of failure of one phase, other 2 can still work.

POWER DISTRIBUTION ON BOARD

The Power Distributed on board a ship needs to be supplied efficiently throughout the ship. For this the power distribution system of the ship is used.

Busbas inspection and maintenance

A shipboard distribution system consists of different component for distribution and safe operation of the system. They are:

- Ship Generator consisting of prime mover and alternator
- Main switch board which is a metal enclosure taking power from the diesel generator and supplying it to different machinery.
- Bus Bars which acts as a carrier and allow transfer of load from one point to another. Circuit breakers which act as a switch and in unsafe condition can be tripped to avoid breakdown and accidents. Fuses as safety device for machinery.
- Transformers to step up or step down the voltage. When supply is to be given to the lighting system a step down transformer is used in the distribution system.
- In a power distribution system, the voltage at which the system works is usually 440v.
- There are some large installations where the voltage is as high as 6600v.
- Power is supplied through circuit breakers to large auxiliary machinery at high voltage.
- For smaller supply fuse and miniature circuit breakers are used.

- The distribution system is three wires and can be neutrally insulated or earthed.
- Insulated system is more preferred as compare to earthed system because during an earth fault essential machinery such as steering gear can be lost.

EMERGENCY POWER

In case of the failure of the main power generation system on the ship, an emergency power system or a standby system is also present. The emergency power supply ensures that thc cssential machinery and system continues to operate the ship.

Emergency power can be supplied by batteries or an emergency generator or even both systems can be used. Rating of the emergency power supply should be made in such a way that it provides supply to the essential systems of the ship such as

a) Steering gear system
b) Emergency bilge and fire p/p
c) Watertight doors.
d) Fire fighting system.
e) Ships navigation lights and emergency lights.
f) Communication and alarm system.

Emergency generator is normally located outside the machinery space of the ship. This is done mainly to avoid those emergency situations wherein access to the engine room is not possible. A switch board in the emergency generator room supplies power to different essential machinery.

SHIP'S ELECTRICAL SYSTEM DESCRIBED

GENERATOR RATING

The generators form the heart of the electrical design and their correct sizing is the key to a safe, workable and economical system. When sizing a marine generator cognisance must be given to the nature of the load. The generator often works on its own and is accordingly susceptible to large system load swings, loads causing distortion, the connection of motors and the connection of large heater elements for air conditioning systems. In addition to satisfying the apparent system load requirements, consideration must be given to the special requirements of any large loads, unusual operational requirements, spare capacity requirements and the required system operating philosophy.

International maritime regulations (e.g. SOLAS), require at least two generators for a ship's main electrical power system. The generators are normally driven from their own dedicated diesel engine but this can be expensive, taking up additional space that could be used for other purposes. For ships engaged on long sea voyages, it can be economical to drive the generators from the main propulsion plant. International maritime regulations also require at least one electrical generator to be independent of the speed and rotation of the main propellers and associated shafting and accordingly at least one generator must have its own prime mover.

If a minimum of two generators is provided, one of which is driven from the propeller shaft, failure of one of the generators could make the ship non-compliant with the International regulations. For this reason many owners opt to provide three generators. One is used for the normal sea load (e.g. the shaft generator), leaving two available to meet any unusually high loads or to provide security when maneuvering. Alternately, the third is retained as a standby set able to provide power should one set fail in service or require specific maintenance work.

In some applications such as a generator supplying a large SCR type load, the generator rating may be increased well beyond its full load value, in order to account for harmonic heating and the inductive requirements of the SCR devices. DCMT has developed its own software to assist in generator sizing.

MAIN SWITCHBOARD

The main elements of a marine distribution system are the main and emergency switchboards, power panel boards, motor controllers, lighting and small power panel boards. The system is generally designed such that under all normal conditions of operation, power is distributed from the main switchboard. The distribution system is designed to keep cable costs to a minimum by distributing to power panels located close to the user services.

The main switchboard is generally located near the centre of the distribution system and this is normally the main engine room or machinery control room. These locations are normally below the ship's waterline or below the uppermost continuous deck of the ship i.e. the bulkhead or main deck. Consequently, in the event of a fire or flooding it is likely that the main generators and switchboard would be disabled. To ensure that electrical supplies are available to emergency and safety systems, an emergency generator and associated emergency switchboard will be located above the main deck in a separate space, completely isolated from the main machinery spaces.

For shipboard installations specific protective systems are required to shut down all ventilation systems and all fuel oil systems in the event of fire. When motor auxiliaries are grouped together and supplied from a motor control center or a grouped distribution panel, this can best be achieved by providing the MCC supply feeder circuit breaker with an undervoltage tripping device and connecting this to the ventilation or fuel systems trip unit. When grouped MCC's or grouped distribution panels are not used, separate cables must be installed for each motor controller. This leads to increased cable costs and increases the systems proness to failure.

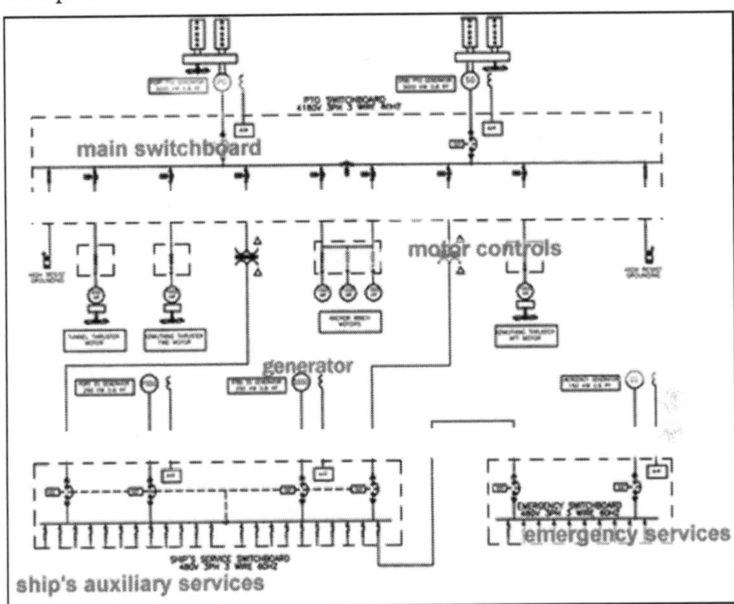

MOTOR CONTROLS

It is often convenient to group motor driven auxiliaries according to their function, e.g. fuel and lubrication oil services, accommodation ventilation systems, machinery ventilation systems, and domestic service systems. The auxiliary motors would be supplied from grouped motor controllers located either in the engine room, in a machinery control room or in a convenient location close to the auxiliary motors. This can often simplify the machinery control functions and required protection systems.

On small ships, e.g. tugs, etc., such grouping is not economical and the major ship's auxiliaries are normally supplied directly from the main switchboard. In this case the motors would be provided with individual starters located adjacent to the motor. For high speed vessels where weight is important, minimum cable weight may be achieved using a "non-distributed" distribution scheme.

Auxiliary motor controls should be arranged in consideration of the general control philosophy applied to the machinery control systems. For ship's that do not have automated machinery operation, the most economic method of control is to provide local starters for each auxiliary motor supplied from power panels located in the same or adjacent spaces. These motors would be manually controlled (start and stopped), locally at the motor's controller (starter). This arrangement minimizes cable costs.

When a centralized machinery control system is required, cables for the motor control functions can be installed back to the machinery control room and the starter push buttons located on a centralized machinery control console. Alternatively, the motors may be grouped together on motor control centres located inside the control room. The motor control functions can then be left on the motor's starter at the MCC or again wired back to a central control desk.

When hard-wired systems are used, the installation is prone to mechanical problems which may cause loose or broken connections and the marine environment which may cause corroded connections. These problems can be eliminated somewhat by using micro-processors and digital control systems.

When fully automatic machinery control is required, these techniques are now in common use and micro-processor devices control the ship's machinery through video display units located in the machinery control room or on the bridge. The ship's auxiliaries are generally controlled with programmable logic controllers (plc's) installed inside the motor control centres and linked through a data bus to the machinery control location. When this type of system is used, the motor control centres can be located either together in the machinery control room or alternatively, distributed throughout the ship close to the motors being controlled. There is little difference in the cabling requirements of either method, however when motor control centers are located outside a dry, atmosphere controlled space such as the machinery control room, a higher degree of mechanical enclosure is required (IP 44 instead of IP 22) and consequently adds to the MCC costs.

EMERGENCY SERVICES

Emergency services would be supplied from the emergency switchboard using distributed panels for navigation, safety and emergency lighting services. These distribution panels are also generally arranged to be above the bulkhead deck. For lighting it is important to ensure that a fire or flooding in one area will not cause loss of lighting in other areas or along escape routes and circuitry must be designed in consideration of the ships general arrangements.

SHIP'S AUXILIARY SERVICES

DCMT's principle design documents for the ships auxiliary services include a load list, load analysis and short-circuit current analysis. In consultation with the client all electrical services on the vessel are identified. Approximate horse-power or kilowatt ratings are obtained for motors. Lighting loads are estimated from the ship's general arrangements and electronic aids are obtained from similar vessels, and a complete load list compiled.

The electrical load analysis uses the load list in order to estimate the expected power demand of the electrical system under specific ship operating conditions. Typical operating conditions would be with the ship, "in transit," "at anchor," "maneuvering," etc. For special vessels, other operating conditions would be appropriate such as "towing" for a tug, "drilling" for a drill ship. The load analysis calculates the expected power demand by multiplying each service power by a "demand" factor. The demand factor is a combined load factor and diversity factor and is the ratio of the estimated power consumption of a service to its normal full load power consumption. The demand factor is determined by an experienced assessment of the estimated power during a four to six hour period when loads may be at their maximum utilization.

DCMT's load analysis obtains load information from the load list. For each service, data banks are searched to determine the service full load current and power factor dependent upon motor operating voltage. This information is used to compute the services' kilowatt and kilovar demand from which is computed the kilovoltamps. By applying the demand factor to each load kW and kvar's and summing all loads for specific operating conditions, the expected generator kilowatts, kilovoltamps and power factor can be computed. By comparing the expected load for the different ship operating conditions, the number and rating of the main generators can be assessed.

Preliminary short-circuit current calculations can be completed once the load analysis and number and rating of generators have been determined. The principle purpose of the short-circuit current calculation is to ascertain the short-circuit rating of the systems protective devices. DCMT has developed several types of short-circuit current calculations which are applied under different circumstances at various stages of the design process. The major contributors to short-circuit current are the generators and motors. Cables and transformers act to reduce the short-circuit current load at a specific location. The most simple short-circuit current analysis is based on an assumed value of the generator's sub-transient reactance and an approximate estimate of the worst case motor loading can be obtained from the load analysis.

The "second stage" short-circuit current analysis is completed when the electrical system conceptual one-line diagram is finished. For this calculation actual subtransient data is used together with cable transformers and other system parameters. This calculation generally results in lower values of short-circuit current. When complete system information is available a "third-stage" short-circuit analysis is completed. This is the most accurate calculation DCMT completes. The calculation determines the decrements of the short-circuit current over a 3 and 5 cycle period.

PRESERVATION OF IRON SHIPS IN THE MARINE ENVIRONMENT

Long-term preservation is a difficult objective for anything that resides in an exposed environment, and this is particularly true for ships in the marine environment. Wooden ships are often considered the most problematic, being of materials that are intended by Mother Nature to rapidly degrade, and to which task she dispatches legions of insects, funguses, bacteria, and marine animals. Yet iron ships, while not subject to such organic onslaught, are hardly less ephemeral. To understand why, we need to look at the material, iron, in both a general and specific sense, and to consider the realities inherent in any preservation effort.

When man produces iron to build a ship he is taking a relatively stable substance - iron ore - and purifying it until it becomes an unstable substance, so unstable in fact that it wants to decompose in the presence of oxygen. A nearly omnipresent element, oxygen is hard to get away from on this planet (some ships have done so, but only by great misfortune) and, as if that is not bad enough, these creations are then tossed into one of the most corrosive environments there is - sea water.

Of course the men who built these ships were not overly concerned with this — they were intent on transporting cargoes or fighting battles on the high seas. They valued the superior strength of iron and probably expected no more than a 20 to 30-year life cycle for their creations. But this is the reality we face as preservationists today, given a mandate to preserve "in perpetuity" objects that were never expected to significantly outlive their creators. Despite a generally high quality of initial construction, most historic ships built of iron or steel have already suffered appreciable deterioration, and many of these have yet to reach the 100th anniversary of their launching — relative youngsters.

THE NATURE OF THE PROBLEM

The historic ships under discussion here range from iron sailing ships of the last century to the welded steel ships of the W.W.II era. The vessels that are the most troublesome and idiosyncratic are the riveted ships of the 19th and early 20th centuries. For one thing, these ships often come into the hands of their preservation-minded caretakers only after suffering years of abuse or neglect (the Balclutha, Eppleton Hall, Wavertree, and Great Britain are among the many). Their partially deteriorated state, in itself, makes them more difficult to maintain. This is in contrast to the more recent military ships that were often (though not always) "mothballed" and carefully preserved for future use. The earlier ships are also more subject to loss of historic integrity, as their forged and riveted structures are harder to repair in kind. The result is that, over time, repeated repairs using alternative methods will alter the original character of the vessel. The following discussion primarily focuses on these ships, although most of the preservation problems, and possible cures, will apply to all iron and steel ships, whether riveted or welded.

THE ENEMY: CORROSION

The old saying "corrosion never sleeps" is definitely true, but may be a little understated; corrosion actually appears to go on binges. Coatings are the primary line of defense against corrosion; they work to keep oxygen and moisture away from the metal. Fortunately we are living in the golden age of protective coatings; zinc-based primers and high-build epoxies have vastly improved the effectiveness, and extended the life span, of anti-corrosive coatings. Perhaps more importantly, many of the new coatings are "surface tolerant", meaning they can be applied over a less-than-well prepared surface. In some cases, this can eliminate the headache of having to sandblast. A detailed discussion of coatings is beyond the scope of this paper. Suffice to say that one should stay abreast of the advances being madc in the field, either by contacting the technical staff of the major paint companies specializing in marine coatings, or by contacting a professional trade organization, such as the Steel Structures Painting Council.

THE UNDERWATER HULL

The portion of an iron ship most vulnerable to corrosion is the underwater hull, and this is where corrosion can have the most catastrophic effect. Along with coatings, an effective weapon for preserving the underwater hull is cathodic protection. This is another area where major advances have been made — the traditional system of using sacrificial anodes has been largely replaced with the "impressed current" system, which pumps electrical current into the hull, turning it into a giant cathode and thereby preventing corrosion. Although impressed current systems have been around for many years, recent advances have been made in the automatic control systems that regulate them. The systems are now fairly reliable and easy to maintain. They will reduce corrosion to a minimum and significantly extend the period between maintenance dry dockings. Any ships afloat in sea water should be under the protection of an impressed current system.

TOPSIDES

The impressed current system will not, however, protect a ship's topsides, which must continue to rely on anti-corrosive coatings. The area of greatest concern is the wind-and-water line, where the combination of wind and splashing sea water invariably causes severe corrosion and wastage (this is often the thinnest area of a hull and where leaks usually begin). The best and thickest epoxy coating is called for here, and will effectively protect the wind-and-water line unless it is mechanically damaged by ice, flotsam, or contact with a pier or another vessel. Measures should be taken to prevent such damage to the waterline area.

BILGES

Bilges are a common corrosion problem because they remain damp. If sloshing bilge water is present, the lower interior of the hull can suffer corrosion rates as severe as those of the wind-and-water line. Bilge water should be kept to an absolute minimum, and bilge areas ventilated to remove moisture. Even with these measures, getting paint to stick to bilge surfaces can be difficult. Some areas may call for a soft-film coating. Once applied, these coatings will make all surfaces extremely slippery. This type of coating is best used in areas that people are not expect to access frequently, such as engine beds or ballast tanks.

Many builders of iron vessels in the late 19th or early 20th century poured cement in the bilges as a protective coating. Compared to the best paint available at the time, cement served far better in protecting hulls from the corrosive effects of sloshing bilge water. Over time though, the system can fail as water gets in through fractures in the cement or along the thin edges where the cement terminates, usually at the turn-of-the-bilge. Moisture starts the corrosion process, and rust expansion lifts the cement in a vicious cycle that eventually spreads throughout the cemented area. This can lead to severe wastage of shell plating that, coupled with exterior wastage, can hole the hull. It may be difficult to tell whether this process is occurring - sounding the cement with a hammer will, in some cases, produce a hollow sound, indicating the cement has lifted; in other cases, particularly where the cement is thick, the cement may sound solid while active corrosion is taking place beneath. Once water has gotten beneath the cement, it never dries out, ensuring that the corrosion will continue. About the only sure way of determining the condition of shell plating beneath cement is to remove the cement.

GALVANIC ACTION

The term "galvanic action" is often used to describe a particular form of corrosion that takes place when dissimilar metals are placed in contact with each other, particularly in the presence of sea water. The result is wastage of the less noble metal. Due to minor dissimilarities between the composition of plates and rivets, galvanic action is often seen in the shell plating of riveted hulls that are afloat in sea water. This generally results in wasting of rivets, which tend to be less noble than the surrounding plate. This form of galvanic action is slow acting, but can result in severe wastage over time. One solution is to encapsulate the rivets and surrounding plate with a thick coating, such as a high-build epoxy, that prevents sea water contact, another is to employ the aforementioned cathodic protection system. Electrolysis

Electrolysis occurs due to ground faults in a ship's electrical system. It can also occur due to stray currents from electrical sources near the ship. In either case, as current passes through a ship's hull, it takes some metal with it, resulting in pitting of the hull exterior similar to that caused by galvanic corrosion. With most permanently moored historic ships, electrolysis results from ground faults in the shore power system. Finding all the ground faults in a ship's electrical system can be daunting. An alternate solution is to install an "isolation transformer" in the electrical shore power line. This effectively keeps the faulty current from passing through the hull as it tries to get back to ground ashore, and does not require that all ground faults are corrected. Stray current from outside sources are more difficult to detect and correct, although the effects can be largely mitigated by the use of an impressed current cathodic protection system.

RUST EXPANSION

One of the weaknesses of riveted construction is the vulnerability of plate seams to corrosion. The joints between plates and other structural members cannot be completely sealed from moisture, even in shell plates that have been caulked. The result is that, over time, corrosion can lead to expanding rust scale that draws joints apart with tremendous force, shearing rivets or pulling the heads though the plate. This can present a problem for the surveyor, as it is often difficult to tell whether rivets still have attachment. For the preservationist, removing rust scale from joints can be difficult, and sealing off the joints nearly impossible. About the only lasting solution, although a costly one, is to remove the rivets, clean out and coat the seam, and re-rivet the joint. A hermetic seal, such as the thick epoxy coatings mentioned earlier, can then be used to seal off the seam

WOOD DECKS ON IRON SHIPS

Planked weather decks are a feature of most early iron ships. An ongoing conflict often exists between the maintenance needs of wooden decks and those of the surrounding ship — wood decks are traditionally washed down with sea water to take advantage of the fungicidal properties of salt. This is robbing Peter to pay Paul in a big way because these same chlorides invariably result in corrosion of adjacent iron structures (hatch coamings, bulwarks, fittings). Wherever possible, an alternative to sea water should be used for deck maintenance. Unfortunately, some of the most effective wood preservatives are also toxic to humans. One safe alternative is sodium borate (sold as TIMBOR (R), a water-soluble preservative that is not corrosive to ferrous metals. Like any water-soluble product, sodium borate will tend to leach out of the wood over time. It therefore must be applied in an ongoing maintenance program in order to be effective.

As every wood deck that was ever laid upon a ship eventually leaks, corrosion is often found on the underlying plate or deck beams. There are few good solutions to this problem and once the corrosion has pitted the tops of beams or plates, the steel-to-wood faying surfaces become irregular and hard to seal. One solution that appears to work is the use of a soft film coating to bed the planking — Eureka Fluid, a brand of lanolin-based anti-corrosive coating, was used in this manner during deck repairs on the ship Balclutha in the 1960s, and again in the 1980s, and appears to have been successful.

MONITORING THE RATE OF CORROSION

To determine the degree of success of our preservation efforts, the rate of corrosion and attendant wastage of a vessel should be monitored closely. Visual inspections should be carried out on a routine basis, with particular attention to some of the problem areas noted above. Measuring shell plate thickness is probably the single most important method for establishing the overall condition and rate of deterioration of an iron hull. The tool of choice for this is the ultrasonic caliper, which measures the thickness of metal by sending an ultrasonic signal though the material. These tools have become cheaper and more sophisticated in recent years and would be a useful addition to the maintenance tool kit of any large vessel. The process of ultrasonic testing (UT) is straightforward, but difficulties are sometimes encountered when attempting to measure some types of malleable iron plate due to imbedded layers of silicone slag. These layers tend to bounce the signal back, thereby giving only a partial reading. An experienced operator can learn to accurately decipher these signals and some newer UT gauges can be calibrated to overcome this problem entirely.

ENOUGH IS ENOUGH

If we are to be successful in preserving iron ships, we must accept that even a minor degree of corrosion is too much. A one-percent annual corrosion rate would perhaps have been acceptable to most ship owners a century ago, but will not achieve the goal of long-term preservation if we are looking beyond the next one-hundred years. Of course, the idea of preserving something "in perpetuity" is patently absurd unless we acknowledge the necessity of constant renewal. Yet, one of the principle values we place in these ships is the craftsmanship they exhibit, something that is not easily reproducible. Who can say what future technologies will bring to the field of maritime preservation — perhaps a solution will be found for this dilemma, but until then, we should do all that we can to not loose our grip on what we have.

ELECTRICAL INSTALLATIONS

GENERAL REQUIREMENTS

General

For all non-HSC craft of more than 500 tons GT and non-HSC passenger craft of less than 500 tons GT (having 'LC' or 'HSLC' notation). Whilst this chapter applies to the electrical engineering equipment and systems on High Speed Craft and Light Crafts intended to be classed, attention should also be given to any relevant statutory regulations of the National Authority of the country in which the craft is to be registered. Electrical services required to maintain the craft in a normal seagoing, operational and habitable condition are to be capable of being maintained without recourse to the emergency source of electrical power.

Electrical services essential for safety are to be maintained under various emergency conditions. The safety of passengers, crew and craft from electrical hazards is to be ensured. Failure mode and effects analysis of the craft is to include the electrical system, taking into account the effects of electrical failure on the systems being supplied. In cases where faults can occur without being detected during routine checks on the installation the analysis is to take into account the possibility of faults occurring simultaneously or consecutively.

DEFINITIONS

Normal operational and habitable condition is condition under which the craft as a whole, the machinery, services, means and aids ensuring propulsion, ability to steer, safe navigation, fire and flooding safety, internal and external communications and signals, means of escape, and emergency boat winches, as well as designed comfortable conditions of habitability are in working order and functioning normally.

Emergency condition is a condition under which any services needed for normal operational and habitable conditions are not in a working order due to the failure of the main source of electrical power. Main source of electrical power is a source intended to supply electrical power to the main switchboard for distribution to all services necessary for maintaining the ship in normal operational and habitable conditions.

Dead craft condition and blackout are both understood to mean a condition under which the main propulsion plant, boilers and auxiliaries are not in operation and in restoring the propulsion, no stored energy for starting the propulsion plant, the main source of electrical power and other essential auxiliaries is to be assumed available. It is assumed that means are available at all times to start the emergency generator.

Main generating station is the space in which the main source of electrical power is situated. Main switchboard is a switchboard which is directly supplied by the main source of electrical power and is intended to distribute electrical energy to craft's services. Emergency switchboard is a switchboard which in the event of failure of the main electrical power supply system is directly supplied by the emergency source of electrical power or transitional source of emergency power and is intended to distribute electrical energy to the emergency services.

Emergency source of electrical power is a source of electrical power, intended to supply the emergency switchboard in the event of failure of the supply from the main source of electrical power. Special category spaces are those enclosed spaces above or below the bulkhead deck intended for the carriage of motor vehicles with fuel, for their own propulsion, in their tanks, into and from which such vehicles can be driven and to which passengers have access.

PLANS

The following plans and details for electrical installation are to be submitted in triplicate for approval.

Single line diagram of main and emergency power and lighting systems which is to include:

a) Ratings of machines, transformers, batteries and semi-conductor converters;
b) All feeders connected to the main and emergency switchboards;
c) Section boards and distribution boards;
d) Insulation type, size and current loadings of cables;
e) Make, type and rating of circuit breakers and fuses.

Simplified diagrams of generator circuits, inter-connector circuits and feeder circuits showing:

a) Protective devices e.g. short circuit, overload, reverse power protection;
b) Instrumentation and synchronizing devices;
c) Preference tripping;
d) Remote stops;
e) Earth fault indication/protection.

Calculations of prospective short circuit currents of main busbar and secondary side of transformers (Additionally load schedule of normal and emergency operating loads on the system estimated for the different operating conditions expected is to be submitted for information). For battery installations, arrangement plans and calculation.

Details of electrically operated passenger and crew safety systems which are to include single line diagrams and a general arrangement plan of the craft showing the vertical fire zones and location of equipment and cable routes of:

a) Emergency lighting;
b) Accommodation fire detection, alarm and extinction systems;
c) Public address system;
d) General alarm;
e) Watertight doors, shell doors and other electrically operated closing appliances.

General Arrangement plan of the craft showing location of major items of electrical equipment, i.e. main and emergency generators, main and emergency switchboards, emergency batteries, motors for emergency services. Arrangement plans of main and emergency switch board and section boarding.

SURVEYS

Electrical machinery and auxiliary services essential for the safety of the craft are to be installed in accordance with the relevant requirements of this chapter, surveyed and have tests witnessed by the Surveyors. Generators, motors and transformers of 100 kW or over intended for essential services are to be surveyed during manufacture and testing.

ESSENTIAL AND OTHER SERVICES

Essential services wheresoever mentioned in the Rules are those which need to be in continuous operation for maintaining the craft's maneuverability with regard to propulsion and steering and for the safety of the craft. Examples of essential services are:

- Steering gear,
- Pumps for variable pitch propellers,
- Fuel oil supply pumps,
- Fuel valve cooling pumps,
- Lubricating oil pumps,
- Circulating and cooling water pumps,
- Automatic sprinkler systems,
- Main engine remote control devices,
- Fire detection and alarm system,
- Communication systems,
- Windlasses,
- Ventilating fans for engine room,
- Air compressors for starting and maneuvering essential mains and auxiliary machinery,
- Air pumps,
- Ballast and bilge pumps,
- Fire pumps,
- Main lighting system for those parts of craft normally accessible to and used by personnel and passengers,
- Navigational aids where required by statutory regulations,
- Navigational lights and special purpose lights where required by statutory regulations,
- Watertight doors and other electrically operated closing appliances,
- Scavenge blowers,
- Valves which are required to be remotely operated,
- Power sources and supply system for supplying the above services.

Services such as following are considered necessary for minimum comfortable conditions of habitability:

- Cooking;
- Heating;

- Domestic refrigeration;
- Mechanical ventilation;
- Sanitary and fresh water.

To maintain the craft in a normal seagoing operational and habitable condition:

- Cargo handling and cargo gear equipment;
- Hotel services, other than those required for habitable conditions;
- Thrusters systems for manoeuvring.

DESIGN AND CONSTRUCTION

Equipment for services essential for the safety of the craft are to be constructed in accordance with the relevant requirements of this chapter. The design and installation of other equipment is to be such that risk of fire due to its failure is minimized. It is to, as a minimum, comply with a National or International Standard revised where necessary for ambient conditions.

VOLTAGE AND FREQUENCY VARIATION

Item	Variations	
	Permanet	Transient
Frequeny	± 5%	± 10% (recovery time 5 seconds)
Voltage	+ 6% - 10%	± 20% (recovery time 1.5 seconds)

Note: The above values do not apply to battery systems.

All electrical appliances supplied from the main or emergency systems are to be so designed and manufactured that they are capable of operating satisfactorily under the normally occurring variations in voltage and frequency. Unless otherwise stated, all equipments are to operate satisfactorily with the variations from its rated value shown in the following table:

INCLINATION OF CRAFT

Emergency and essential electrical equipment is to operate satisfactorily under the conditions as shown in Table below.

Table. Machinery inclination.

Installations/Components	Angle of inclination [1)]			
	Athwartships		Fore-and-aft	
	Static	Dynamic	Static	Dynamic
Main and auxiliary machinery	15°	22.5°	5°	7.5°
Safety equipment e.g. emergency power installations, emergency fire pumps and their devices switchgear, electrical and electronic appliances[2)] and remote control systems	22.5°	22.5°	10°	10°

Notes: 1) Athwartships and core-and-aft inclinations occur simultaneously.
2) Switches and controls are to remain in their last set position.

LOCATION AND CONSTRUCTION

Electrical equipment is, as far as is practicable, to be accessibly placed, clear of flammable material in well ventilated, adequately lighted spaces, in which flammable gases cannot accumulate and where it is not exposed to risk of mechanical injury or damage from water, steam or oil. Where necessarily exposed to such risks, the equipment is to be suitably constructed or enclosed. Live parts are to be guarded where necessary.

All electrical apparatus is to be so constructed and so installed that it does not cause injury when handled or touched in the normal manner. Insulating materials and insulated windings are to be flame retardant and resistant to tracking, moisture, sea air and oil vapour unless special precautions are taken to protect them. Equipment is not to remain alive through the control circuits and/or pilot lamps when switched off by the control switch.

This does not apply to synchronizing switches and/or plugs. The operation of all electrical equipment and the lubrication arrangements are to be efficient under such conditions of vibration and shock as arise in normal practice. All nuts and screws used in connection with current-carrying and working parts are to be effectively locked. Conductors and equipment are to be placed at such a distance from the magnetic compasses, or are to be so disposed, that the interfering magnetic field is negligible when circuits are switched on and off.

EARTHING

All accessible metal of the electric installation, other than current carrying accessible parts should be earthed.

The following parts may be exempted from the requirements:

a) Lamp-caps, where suitably shrouded;
b) Shades, reflectors and guards supported on lamp-holders or light fittings constructed of, or shrouded in, non-conducting material;
c) Metal parts on, or screws in or through, non-conducting materials, which are separated by such material from current-carrying parts and from earthed non-current carrying parts in such a way that in normal use they cannot become live or come into contact with earthed parts;
d) Apparatus which is constructed in accordance with the principle of double insulation;
e) Bearing housings which are insulated in order to prevent circulation of current in the bearings;
f) Clips for fluorescent lamps;
g) Cable clips and short lengths of pipes for cable protection;
h) Apparatus supplied at a voltage not exceeding 55 V direct current or 55 V, root mean square, between conductors, or between any

conductor and earth in a circuit isolated from the supply. Autotransformers are not to be used for the purpose of achieving the alternating current voltage;

i) Apparatus or parts of apparatus which although not shrouded in insulating material is nevertheless otherwise so guarded that it cannot be touched and cannot come in contact with exposed metal.

Metal coverings of cables are to be effectively earthed at both ends of the cable. In final sub-circuits, other than those installed in hazardous zones or spaces, earthing at the supply end only will be considered adequate. Single point earthing may be accepted for instrumentation cables if desirable for technical reasons.

The electrical continuity of all metal coverings of cables throughout the length of the cable, particularly at joints and tappings, is to be ensured. Metal parts of portable appliances, other than current-carrying parts and parts are to be earthed by means of an earth-continuity conductor in the flexible cable or cord through the associated plug and socket-outlet.

Earthing conductors are to be of copper or other corrosion-resistant material and be securely installed and protected where necessary against damage and also, where necessary, against electrolytic corrosion. Connections are to be so secured that they cannot work loose under vibration. The nominal cross-section areas of copper earthing conductors are, in general to be equal to the cross section of the current carrying conductor upto 16 mm^2.

Above this figure they are to be equal to at least half the cross-section of the current-carrying conductor with a minimum of 16 mm^2. Every other earthing conductor is to have a conductance not less than that specified for an equivalent copper earthing conductor.

The connection of the earthing conductor to the hull of the craft is to be made in an accessible position and is to be secured by a screw or stud of diameter not less than 6 mm which is to be used for this purpose only. Bright metallic surfaces at the contact areas are to be ensured immediately before the nut or screw is tightened and, where necessary, the joint is to be protected against electrolytic corrosion. The connection is to remain unpainted.

ELECTRICAL BONDING FOR THE CONTROL OF STATIC ELECTRICITY

In non-metallic craft, all metallic parts of the craft are to be electrically bonded together, as far as possible, in consideration of galvanic corrosion between dissimilar metals, to ensure an earth return path and to connect the craft to the water when water-borne.

This does not apply to isolated components, which cannot become live, nor require control of static electricity.

Bonding straps for the control of static electricity are required for piping systems, including pressure refuelling points, which are not electrically continuous throughout their length and for flammable products, which are not permanently connected to the hull of the craft either directly or via their bolted or welded supports and where the resistance between them and the hull exceeds 1MW. Where bonding straps are required for the control of static electricity, they are to be robust, that is, having a cross sectional area of about 10 mm^2. Where bonding conductors are provided for lightning discharge currents, they are to have a minimum cross section of 50 mm^2 in copper.

ENCLOSURES

The enclosure types given in Table below are required as a minimum.

Table. Enclosure types in relation to location

(N: Normally not accepted for installation in this location).

Location	Switch-boards	Control gear and motor starters	Rotating machines	Trans-formers and rectifiers	Lighting fittings	Heating applian-ces	Socket outlets	Accesso-ries such as switches, connecti-on boxes
Engine rooms								
Above the floor	IP 22	IP 22	IP 22	IP 22	IP 22	IP 22	IP 44	IP 44
Dry control rooms	IP 22	IP 22	IP 22	IP 22	IP 22	IP 22	IP 22	IP 22
Below the floor	N	N	IP 44	N	IP 44	IP 44	N	IP 44
Closed compartments for fuel oil and lub.oil separators	IP 44	IP 44	IP 44	IP 44	IP 44	IP 44	N	IP 44
Fuel oil tanks	N	N	N	N	N	N	N	N
Ballast and other water tanks, bilge wells	N	N	IP 68	N	N	IP 68	N	N
Ventilation ducts	N	N	IP 44	N	N	N	N	N
Deckhouses, forecastle spaces, steering gear compartments and similar spaces	IP 22	IP 22	IP 22	IP 22	IP 22	IP 22	IP 44	IP 44
Ballast pump rooms and similar rooms below the loadline	IP 44	IP 44	IP 44	IP 44	IP 44	IP 44	IP 56	IP 56
Cargo holds	N	N	IP 44	N	IP 55	N	IP 56	IP 56
Open deck, keel ducts	IP 56	IP 56	IP 56	IP 56	IP 55	IP 56	IP 56	IP 56
Battery rooms, lamp rooms, paint stores, stores for welding gas bottles	N	N	EX	N	EX	EX	N	EX
Dry accommodation spaces	IP 20	IP 20	IP 20	IP 20	IP 20	IP 20	IP 20	IP 20
Bath rooms and showers	N	N	N	N	IP 44	IP 44	N	IP 56
Galleys, laundries and similar rooms	IP 44	IP 44	IP 44	IP 44	IP 44	IP 44	IP 44	IP 44

CREEPAGE AND CLEARANCE DISTANCES

Distance between live parts and between live parts and earthed metal, whether across surfaces or in air, are to be adequate for the working voltages considering the nature of the insulating material and the transient over voltages developed by switch and fault conditions.

For bare busbars the minimum clearance distances are to be observed. Where necessary these distances are to be increased to allow for the electromagnetic forces involved and also depend upon insulating material, dust, deposits, humidity etc.

ADDITIONS OR ALTERATIONS

No addition, temporary or permanent, is to be made to the approved load of an existing installation until it has been ascertained that the current carrying capacity and the condition of the existing equipment including cables and switchgear are adequate for the increased load.

Plans are to be submitted for consideration and the alterations or additions are to be carried out under the survey and to the satisfaction of the Surveyors.

Table. Minimum clearances distances.

Voltage between phases or poles	Minimum clearance to earth		Minimum clearance between phases or poles	
	In air [mm]	In oil [mm]	In air [mm]	In oil [mm]
660 or less	16	-	19	-
2200	38	-	38	-
3300	51	13	51	19
6600	63	19	89	25

INTEGRATED ELECTRIC PROPULSION

Integrated electric propulsion (IEP) or full electric propulsion (FEP) or integrated full electric propulsion (IFEP) is an arrangement of marine propulsion systems such that gas turbines or diesel generators or both generate three phaseelectricity which is then used to power electric motors turning either propellers or waterjet impellors. It is a modification of the combined diesel-electric and gas propulsion system for ships which eliminates the need for clutches and reduces or eliminates the need for gearboxes by using electrical transmission rather than mechanical transmission of energy.

Eliminating the mechanical connection between the engines and the propulsion has several advantages including increased freedom of placement of the engines, acoustical decoupling of the engines from the hull which makes the ship less noisy, and a reduction of weight and volume. Reducing acoustic signature is particularly important to naval vessels seeking to avoid detection and to cruise ships seeking to provide passengers with a pleasant voyage, but is of less benefit to cargo ships.

Because ships anyway require electricity even when not underway, having all of the engines produce electricity reduces the number of engines needed compared to more traditional arrangements in which one pool of engines provides electricity and another pool of engines provides propulsion, reducing capital costs and maintenance costs.

A typical integrated electric propulsion arrangement includes both diesel generators and gas turbines. The advantages of gas turbines include much lower weight and smaller size than diesels of similar power, and much less noise and vibration, but they are efficient only at or near maximum power. Diesel generators have the advantage of high efficiency over a wide range of power levels. Using them in combination allows for the benefits of a full range of operational efficiency, a low-vibration quiet mode of operation, and some reduction in weight and volume relative to a diesel-only arrangement. In naval vessels, a pool of diesel generators are typically used to provide a base load and enough power to achieve cruise speed. The gas turbines are used to provide peak power for higher speeds and may be required to operate weapon systems with high power demands. In passenger ships, one or more gas turbines are used for fast cruising. The diesels provide reliable redundancy and an efficient source of electricity when in port, at anchor, or drifting. A diesel-electric system is an integrated electric propulsion system in which no gas turbines are used and all of the engines are diesel. Some yachtsuse only gas turbines for integrated electric propulsion without any diesel engines.

MORE ON PROPULSIVE SYSTEMS

Propulsive system consists of the ship hull and propulsive unit. The propulsive unit is an executive part of the main ship power plant designed for conversion of the mechanical energy produced by the engine into the vessel moving work. The propulsive system consists of:

- A ship hull,
- A main engine;
- A main gear;
- A shaft line;
- A propulsion device.

The components of the propulsive unit are selected with a glance to the efficiency of the main ship power plant being the key indicator of the perfection of the whole propulsive system.

The main gear (power transmission) is a complex of elements which transmit torque from the crankshaft of the internal combustion engine or the turbine rotor to the propulsor, total torque of several engines with their subsequent transmission to one shaft, and reverse the shaft train and propeller of irreversible engines. The most popular transmissions are mechanical ones, although other transmissions are also used (electrical, mixed and complex).

The standard transmission of the medium running engine consists of a coupling, reducing gear, and shaft train. With high running internal combustion engines and turbines, double- or triple-reduction gear units are installed. The reducing gear is not used in the propulsive unit with low-running engines. It should be noted that the transmission is necessary in case of the vessel equipped with a single-machine ship power plant, and a fixed pitch propeller used as its propulsor, as well as when the engine torque is insufficient for rotation of a propeller screw with giver frequency. Electrical main gears have a number of benefits, among which are a short shaft train, possibility of use with high-running irreversible internal combustion engines, main SPP employment for general need energy supply, survivability and self-regulation ability. With that electrical transmissions are expensive, low efficient, and, at low rate speed, have big dimensions and weight.

The hydraulic transmission is low efficient, so most frequently it is used in combination with the mechanical one. The shaft train is an assembly of mechanisms, devices and connections used for coupling of the propeller screw with the screw shaft as well as for thrust transmission from the propeller screw to the vessel. It consists of a screw shaft, line shafts, plummer blocks, thrust shaft with bearings and stern gear. Besides, bearings and thrust shafts can lack some designs of shaft trains. And there are designs of the engine-propulsor system without a traditional shaft train.

PROPULSIVE SYSTEM APPLICATION: NEW TRENDS

Propulsive system efficiency upgrading can be achieved by various means – upgrading of efficiency of the propulsor or engine, optimization of the hull shape. Bulb fore end can reduce wave-pattern resistance by 15 or more per cent, however often ship vibration nullifies its effect. Wave-pattern resistance can also be reduced by design of the maximum narrow hull with simultaneous increasing of the ship length.

Recently, some wave-piercing catamarans were built abroad. In addition to a narrow hull, they differ in a breakwater at the forebody. Another way of raising efficiency consists in optimization of the propeller-rudder system shape, aft end design and propulsive unit operation.

Engine-propulsor coordinated interaction is necessary for optimal performance of the propulsive unit. Most frequently, propulsive indicators can be raised via propeller diameter increase with shaft speed down. Application of steerable propellers with single-shaft screws also improved performance. This way, the diameter of steerable propellers could be decreased while raising efficiency by 20%.

Replacement of the large diameter propeller screw with two reverse coaxial propellers is another trend. In such event, since a special reducing gear and hollow screw shaft is necessary, the design of the whole propulsive unit becomes more complicated.

Vessel propulsive properties can also be improved via employment of multi-shaft main gears. Introduction of electric propulsive installations is very promising. Thus, vessels equipped with electric propulsive installations and water-jet propulsors are able to accelerate up to 60 knots. The combination of this system with steerable propellers ensures light mass decreasing by 10-15% and raising engine efficiency up to 98%. At the speed up to 25 knots, water-jet propulsors are inferior to popular propeller screws in respect of propulsive properties, however at the speed of about 40 knots, their properties are comparable. At the speed over 50 knots, water-jet propulsors leave propeller screws behind.

Potential of raising performance of water-jet propulsors is being sought both in our country and abroad. For the most part, such search is focused on decreasing hydraulic loss in the duct and improving water-intake devices. One of the prospective methods to decrease hydraulic loss in the propulsor duct can be modification pipe streamlined surface microgeometry by means of flow direction micro-grooves (riblets).

Domestic professionals offered the standard concentrated water intake through a hull aperture or special-purpose nozzle which is usually used in modern water-jet propulsors to be replaced with so-called distributed shear layer water intake through the transverse edgewise conduit. Propulsive performance can be improved by decrease of hull roughness. Cover plate cockle, weld flatness and dimensions, lacquer coat properties, hull and propeller antifoulant application and so on are of importance in this case.

Neither use of heavy-duty engines, nor modernization of a screw design or optimization of a hull shape ensure significant increasing of the ship speed. In such case, one of the solutions can be installation of propeller screws in the bows and not in the aft. Under certain conditions, this creates so-called following wake which significantly improves propulsive properties of the ship. Control system for propulsive complex performance is also important. Such device contains fuel flow rate, power, revolution, and navigable depth transducers. It also includes engine and propulsor diagnostic, logic processing, data display and individual alarm units.

SHIP STABILITY

Ship stability is an area of naval architecture and ship design that deals with how a ship behaves at sea, both in still water and in waves, whether intact or damaged. Stability calculations focus on the center of gravity, center of buoyancy, and metacenter of vessels and on how these interact.

HISTORY

Ship stability, as it pertains to naval architecture, has been taken into account for hundreds of years. Historically, ship stability calculations for ships relied on rule of thumb calculations, often tied to a specific system of measurement. Some of these very old equations continue to be used in naval architecture books today. However, the advent of calculus-based methods of determining stability, particularly Pierre Bouguer's introduction of the concept of the metacenter in the 1740'S ship model basin allows much more complex analysis.

Fig. A model yacht being tested in thetowing tank of Newcastle University.

Master shipbuilders of the past used a system of adaptive and variant design. Ships were often copied from one generation to the next with only minor changes being made; by replicating a stable design serious problems were not often encountered. Ships today still use the process of adaptation and variation that has been used for hundreds of years; however computational fluid dynamics, ship model testing and a better overall understanding of fluid and ship motions has allowed much more analytical design. Transverse and longitudinal waterproof bulkheads were introduced in ironclad designs between 1860 and the 1880s, anti-collision bulkheads having been made compulsory in British steam merchant ships prior to 1860. Before this a hull breach in any part of a vessel could flood the entire length of the ship. Transverse bulkheads, while expensive, increase the likelihood of ship survival in the event of damage to the hull, by limiting flooding to breached compartments separated by bulkheads from undamaged ones. Longitudinal bulkheads have a similar purpose, but damaged stability effects must be taken into account to eliminate excessive heeling. Today, most ships have means to equalize the water in sections port and starboard (cross flooding), which helps to limit the stresses experienced by the structure and also to alter the heel and/or trim of the ship.

ADD-ON STABILITY SYSTEMS

These systems are designed to reduce the effects of waves or wind gusts. They do not increase the stability of the vessel in a calm sea. The International

Maritime OrganizationInternational Convention on Load Lines does not mention active stability systems as a method of ensuring stability. The hull must be stable without active systems.

PASSIVE SYSTEMS

Bilge Keel

Fig. A bilge keel.

A bilge keel is a long fin of metal, often in a "V" shape, welded along the length of the ship at the turn of the bilge. Bilge keels are employed in pairs (one for each side of the ship). A ship may have more than one bilge keel per side, but this is rare. Bilge keels increase the hydrodynamic resistance when a vessel rolls, thus limiting the amount of roll a vessel has to endure.

Outriggers

Outriggers may be employed on certain vessels to reduce rolling. Rolling is reduced either by the force required to submerge buoyant floats or by hydrodynamic foils. In some cases these outriggers may be of sufficient size to classify the vessel as a trimaran; however on other vessels they may simply be referred to as stabilizers.

Antiroll Tanks

Antiroll tanks are tanks within the vessel fitted with baffles intended to slow the rate of water transfer from the port side of the tank to the starboard side. The tank is designed such that a larger amount of water is trapped on the higher side of the vessel. This is intended to have an effect completely opposite to that of the free surface effect.

Paravanes

Paravanes may be employed by slow-moving vessels (such as fishing vessels) to reduce roll.

ACTIVE SYSTEMS

Many vessels are fitted with active stability systems. Active stability systems are defined by the need to input energy to the system in the form of a pump, hydraulic piston, or electric actuator. These systems include stabilizer fins attached to the side of the vessel or tanks in which fluid is pumped around to counteract the motion of the vessel.

Stabilizer Fins

Active fin stabilizers are normally used to reduce the roll that a vessel experiences while underway or, more recently, while at rest. The fins extend beyond the hull of the vessel below the waterline and alter their angle of attack depending upon heel angle and rate-of-roll of the vessel. They operate similar to airplane ailerons. Cruise ships and yachtsfrequently use this type of stabilizer system. When fins are not retractable, they constitute fixed appendages to the hull, possibly extending the beam or draft envelope, requiring attention for additional hull clearances.

While the typical "active fin" stabilizer will effectively counteract roll for ships underway, some modern active fin systems have been shown capable of reducing roll motion when vessels are not underway. Referred to as zero-speed or Stabilization at Rest, these systems work by moving fins of special design, with the requisite acceleration and impulse timing to create effective roll cancellation energy.

Gyroscopic Internal Stabilizers

Gyroscopes were first used to control a ship's roll in the late 1920s and early 1930s for warships and then passenger liners. The most ambitious use of large gyros to control a ship's roll was on an Italian passenger liner, the SS Conte di Savoia, in which three large Sperry gyros were mounted in the forward part of the ship. While it proved successful in drastically reducing roll in the westbound trips, the system had to be disconnected on the eastbound leg for safety reasons. This was because with a following sea (and the deep slow rolls this generated) the vessel tended to 'hang' with the system turned on, and the inertia it generated made it harder for the vessel to right herself from heavy rolls.

Gyro stabilizers consist of a spinning flywheel and gyroscopic precession that imposes boat-righting torque on the hull structure. The angular momentum of the gyro's flywheel is a measure of the extent to which the flywheel will continue to rotate about its axis unless acted upon by an external torque. The higher the angular momentum, the greater the resisting force of the gyro to external torque (in this case more ability to cancel boat roll).

A gyroscope has three axes: a spin axis, an input axis, and an output axis. The spin axis is the axis about which the flywheel is spinning and is vertical for a boat gyro. The input axis is the axis about which input torques are applied. For a boat, the principal input axis is the longitudinal axis of the boat since that is the axis around which the boat rolls. The principal output axis is the transverse (athwartship) axis about which the gyro rotates or precesses in reaction to an input.

When the boat rolls, the rotation acts as an input to the gyro, causing the gyro to generate rotation around its output axis such that the spin axis rotates to align itself with the input axis. This output rotation is called precession and, in the boat case, the gyro will rotate fore and aft about the output or gimbal axis. Angular momentum is the measure of effectiveness for a gyro stabilizer, analogous to horsepower ratings on a diesel engine or kilowatts on a generator. In specifications for gyro stabilizers, the total angular momentum (moment of inertia multiplied by spin speed) is the key quantity. In modern designs, the output axis torque can be used to control the angle of the stabilizer fins to counteract the roll of the boat so that only a small gyroscope is needed. The idea for gyro controlling a ship's fin stabilizers was first proposed in 1932 by a General Electric scientist, Dr Alexanderson. He proposed a gyro to control the current to the electric motors on the stabilizer fins, with the actuating instructions being generated by thyratron vacuum tubes.

CALCULATED STABILITY CONDITIONS

When a hull is designed, stability calculations are performed for the intact and damaged states of the vessel. Ships are usually designed to slightly exceed the stability requirements (below), as they are usually tested for this by a classification society.

Intact Stability

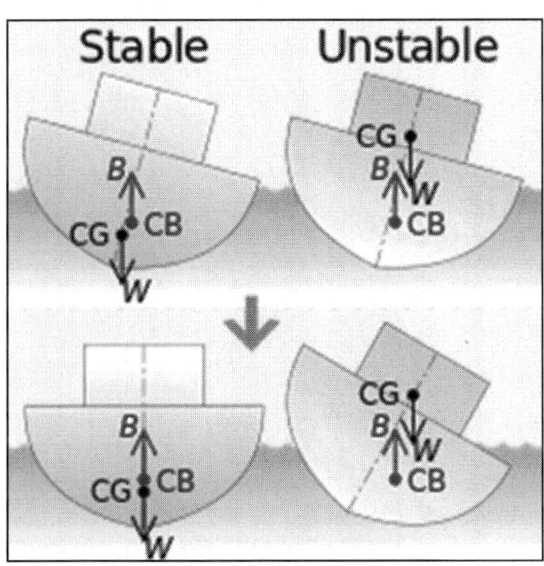

Illustration of the stability of bottom-heavy (left) and top-heavy (right) ships with respect to the positions of their centres of buoyancy (CB) and gravity (CG)

Intact stability calculations are relatively straightforward and involve taking all the centers of mass of objects on the vessel which are then computed/ calculated to identify the center of gravity of the vessel, and the center of buoyancy of the hull. Cargo arrangements and loadings, crane operations, and the design sea states are usually taken into account. The photo at right illustrates a common misconception about vessels. In fact, in the vast majority of vessels, the center of gravity is well above the center of buoyancy.

The ship is stable because as it begins to heel, one side of the hull begins to rise from the water and the other side begins to submerge. This causes the center of buoyancy to shift toward the side that is lower in the water.

The job of the naval architect is to make sure that the center of buoyancy shifts outboard of the center of gravity as the ship heels. A line drawn from the center of buoyancy in a slightly heeled condition vertically will intersect the centerline at a point called the metacenter. As long as the metacenter is further above the keel than the center of gravity, the ship is stable in an upright condition.

Damage Stability (Stability in the Damaged Condition)

Damage stability calculations are much more complicated than intact stability. Software utilizing numerical methods are typically employed because the areas and volumes can quickly become tedious and long to compute using other methods.

The loss of stability from flooding may be due in part to the free surface effect. Water accumulating in the hull usually drains to the bilges, lowering the centre of gravity and actually decreasing (It should read as increasing, since water will add as a bottom weight there by increasing GM) the metacentric height. This assumes the ship remains stationary and upright. However, once the ship is inclined to any degree (a wave strikes it for example), the fluid in the bilge moves to the low side. This results in a list.

Stability is also lost in flooding when, for example, an empty tank is filled with seawater. The lost buoyancy of the tank results in that section of the ship lowering into the water slightly. This creates a list unless the tank is on the centerline of the vessel.

In stability calculations, when a tank is filled, its contents are assumed to be lost and replaced by seawater. If these contents are lighter than seawater, (light oil for example) then buoyancy is lost and the section lowers slightly in the water accordingly.

For merchant vessels, and increasingly for passenger vessels, the damage stability calculations are of a probabilistic nature. That is, instead of assessing the ship for one compartment failure, a situation where two or even up to three compartments are flooded will be assessed as well. This is a concept in which the chance that a compartment is damaged is combined with the consequences for the ship, resulting in a damage stability index number that has to comply with certain regulations.

REQUIRED STABILITY

In order to be acceptable to classification societies such as the Bureau Veritas, American Bureau of Shipping, Lloyd's Register of Ships and Det Norske Veritas, the blueprints of the ship must be provided for independent review by the classification society. Calculations must also be provided which follow a structure outlined in the regulations for the country in which the ship intends to be flagged.

Within this framework different countries establish requirements that must be met. For U.S.-flagged vessels, blueprints and stability calculations are checked against the U.S. Code of Federal Regulations and International Convention for the Safety of Life at Sea conventions. Ships are required to be stable in the conditions to which they are designed for, in both undamaged and damaged states. The extent of damage required to design for is included in the regulations. The assumed hole is calculated as fractions of the length and breadth of the vessel, and is to be placed in the area of the ship where it would cause the most damage to vessel stability.

In addition, United States Coast Guard rules apply to vessels operating in U.S. ports and in U.S. waters. Generally these Coast Guard rules concern a minimum metacentric height or a minimum righting moment. Because different countries may have different requirements for the minimum metacentric height, most ships are now fitted with stability computers that calculate this distance on the fly based on the cargo or crew loading. There are many commercially available computer programs used for this task.

SHIP ELECTRIC EQUIPMENT

Ship electrical facilities is a complex of instruments and devices for energy conversion and transmission. This complex is formed by transformers, transducers, terminal boxes, various cables, wires, electromechanical, electrical and electronic control, monitoring and protective instruments, regulators, lighting and heating devices, etc.

In view of pattern of utilization, ship electric facilities must meet a number of requirements, including small dimensions and weight, increased strength and improved insulation, upgraded vibration, humidity, salt, and water entry protection. All equipment is to meet GOST R 51683-2000.

Generators

AC or DC generators as well as accumulators are used as onboard power systems. Generators could be diesel, shaft-driven, and turbo generators; and by purpose they can be classified into main, harbor, back-up and emergency generators. Diesel-generators consist of a generator itself and diesel, turbine generators – of a turbine and generator. Shaft-driven generators work on power take-off of the ship power plant.

Diesel generators are very efficient, independent and stand-alone, but they have small service life. Steam turbo generator efficiency is less and they need steam pressure maintenance even at no load having, at the same time, smaller weight and dimensions as well as larger service life. Gas turbine generators have advantages both of a steam turbine (small dimensions, weight and large service life) and of diesel generators (independence) but they lack economical operation.

Since shaft-driven generator speed as well as repeatable accuracy of current output characteristics are in direct relation with ship speed, shaft-driven generators are installed at ships which MSPP does not often changes its running regime.

With application of diesel and turbo generators used as auxiliary SPP, their fuel consumption lowers and service life grows. Back-up generators are designated for replacement of disabled main generators. An emergency generator ensures operation of the main ship systems in case of complete failure of the main SPP (i.e. main and back-up generators). Generally, accumulators as main power supplies at small ships, and as emergency and back-up power supplies at large ships. Recently, lithium batteries added to acid and alkaline accumulators.

Transformers

At ships the following transformers are used:

- Potential transformers (for transformation of high voltage into low in relay protection and emergency control and measuring circuits);
- Current transformers (alteration of primary current parameters into necessary value);
- Isolating transformers (power supply safety improving and electrocorrosion prevention);
- mains transformers (energy transformation in installations and mains), etc.

Electric energy transducers deserve special focus. Today, high frequency AC and DC transduction methods are used. Despite complicacy of transduction according such method, use of application-specific integrated circuits ensures not only high reliability of such units but their cost.

Electricity Distribution Plants

Protective, commutation, measuring, and alarm devices installed at special-purpose metallic structures are electricity distribution plants of the ship electrical network. They are used for electrical circuit disconnection and connection, their status check, various parameters measuring, etc. They can be:

- Ship general distribution board;
- Distribution boards (DB);
- Consumer terminal boxes (CB);
- Control panels;
- Control boxes, and
- Special-purpose devices.

Ship electrical network is a network of conducting cables laid to minimize possibility of its failure. Ship cables can havc various sheaths (from rubber and varnished insulating fabric to flux coating). Besides, metallic armor braiding should protect cables form mechanical damage. Cable and its make-up selection depends on numerous conditions, such as sensed current, cable location, etc.

Apart from the above, ship electric installations also include systems and units of:

- Electric protection;
- Diagnostics;
- Control;
- Automatic operation;
- Reactive power compensation;
- Electric power supply of communications-electronics equipment;
- Power filters;
- Current limitation devices;
- Various sockets and inputs;
- Junction boxes;
- Heating and lighting facilities, etc.

Current Situation and Outlooks

Ship electric equipment is permanently improving. Energy quality factors, reliability, large service life as well as weigh and dimensions are relevant for them. Today, module system development ensuring redundancy of main units and parts are on the front burner. This way, reliability is improved as well as repair and current maintenance time is decreased.

LED illuminators are one of the achievements of lighting equipment manufacturers. Many companies manufacture navigation lights on their base. Application of such facilities help save energy (energy consumption of a LED illuminator is by one order lower comparing to traditional incandescent lamp) with the same or greater brightness. Besides, service life of LED illuminators is much more than that of incandescent lamps, they are shock, push and vibration resistant, and can be hermetic. Domestic electric facilities are not bad. Devices meeting the most perfectionist requirements are designed and produced in Russia. Their main shortage is specific consumption of materials which means relatively large weigh and dimensions.

2

Marine Electrical Power Distribution and System

A ship can be considered as a mobile power plant. Most of the machinery is driven electrically, which thus makes for a huge electrical power demand. This makes a shipboard electrical power distribution system very important. This article is about the layout of the main switch board and the bus bar.

INTRODUCTION

Electricity has to be distributed with minimal losses after generation. Unlike a shore based transmission system, where the lengths of the conductors run throughout a country, a shipboard electrical distribution system is short and simple. In this article, the general layout of the main electrical distribution system, along with the main switchboard and emergency switchboard arrangements, will be discussed.

The Main Bus Bar

The main bus bar can be found inside the main switch board. Onboard a merchant ship, the supply is usually 440 volts, 60 Hertz, however higher voltages of 6.6kV are possible on LNG tankers, special purpose ships, and passenger vessels having electrical propulsion. The rating of the main bus bar is decided by the type of ship, the critical machinery on board required for cargo operations, and the machinery required for normal sea service.

If a ship is steam propelled, it might have steam turbine driven alternators and a diesel driven alternator as a back-up. If the propulsion is by electric motors, then the vessel might require huge diesel driven generators which usually operate at 6.6kV.

Ship Type Power Requirement

A 10,000 dwt general cargo ship 1 MW.

Tankers 1.5 to 5 MW.

Containerships (10000 TEU) (3.3kV) 8 MW

A Modern LNG carrier (Q-max/Q-flex) (6.6kV) 12 MW

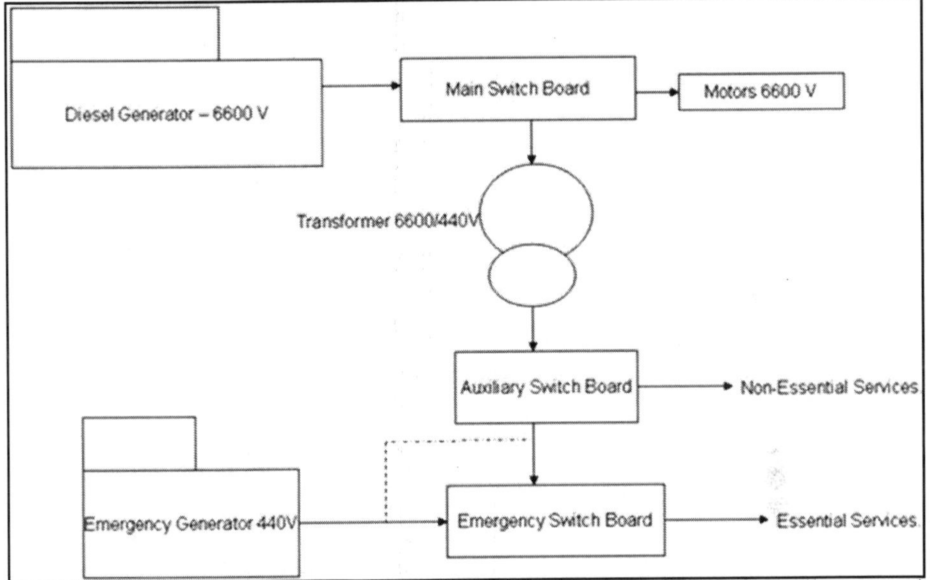

The main bus bar has three heavy, thick bars of conductor (usually copper), running horizontally throughout the length of the main switch board. Each of the conductors is insulated between each other and wherever it is supported or clamped to the switchboard.

The bus bar lies at the bottom most portion of the switchboard such that it can be easily supported and insulated. At times, there may be two divisions of the bus bar which are connected to each other by either a circuit breaker or a clamping (sliding contact) device.

The main switchboard is the main power distribution center of the ship. Thus the main bus bars are contained within the switchboard feeding various ship board auxiliaries.

A ship may contain two or more generators connected to the main bus bar via the circuit breaker. Various protection for the generators like overload, reverse power, etc. are connected to circuit breaker such that the faulty generator is electrically isolated from the main bus bar. From the main bus bar, the electrical power is supplied to various ship board auxiliaries like pumps, blowers, compressors, etc.

The main switch board has various measuring and monitoring devices like ammeters, voltmeters, frequency meters, watt meters, synchroscope, and power factor meters.

A generic layout of a shipboard distribution system is attached here. Refer to the diagram below for the following explanation; however this is just a simplified version of the main power distribution. This may become complex as the size of the vessel and amount of machinery increase in number.

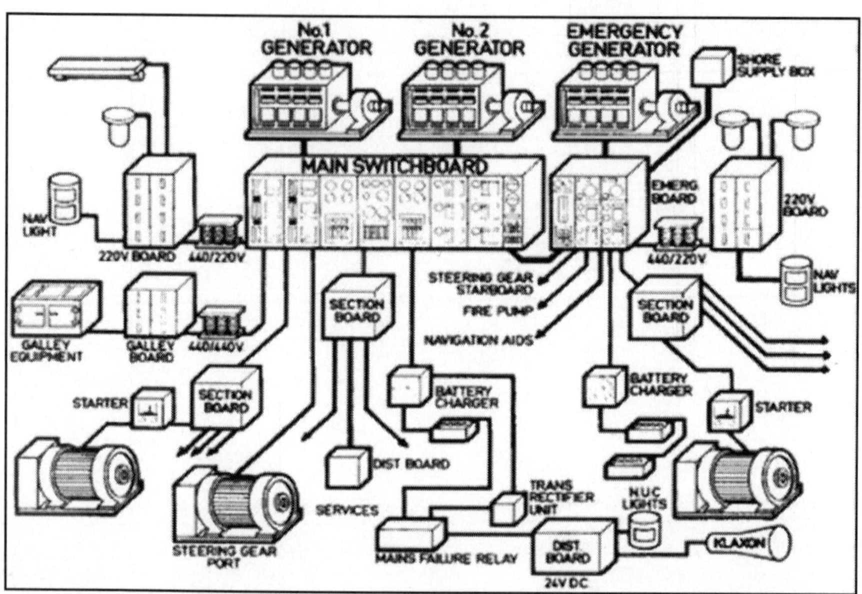

The main switch board gets its supply from the main generators located in the engine room. The distribution system can be divided into Feeder side and the Load side. The feeders are the generators and the loads are various ship board auxiliary machineries. The Load side is further divided into Essential and Non-Essential services. Essential services are very critical auxiliaries which are directly affecting the safety of personnel, ship in terms of navigation and propulsion. They include supply to navigational aids like radars, communication equipment, navigational lights, and steering gear motors. These essential services may be supplied directly from the main switch board or via sectional boards or distribution boards. Non-essential services are those that do not affect the safety of ship and personnel. For example, air conditioning compressors and fans are non-essential services as they don't affect the safety of the ship or personnel.

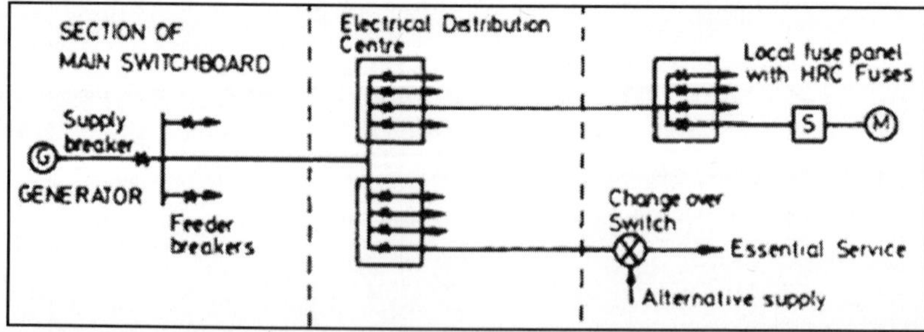

The operating voltage is stepped down to 220 volts and this voltage is given to the 220 Volts auxiliaries' panel. From this panel, the supply goes for various lighting systems and other sockets for accommodation and galley for domestic use.

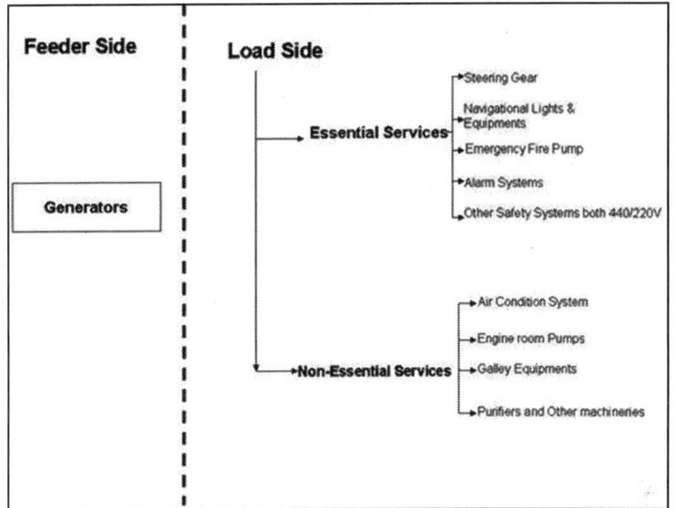

Distribution Boards/Group Start Panels

Shipboard machinery may not be connected to the main switchboard directly. There are small motors and other devices that consume very little power and thus they may be grouped together. Instead of providing cables for every such individual motor, a single cable is taken out from the main switch board, supplying the distribution board which has a small bus bar (equivalent to supply the motors). From this small bus bar, the supply is provided for various starters of these motors via the circuit breakers. In this way, the protection for the machinery is also precise, as a small fault in such small motors will not operate the protection devices on the main switch board.

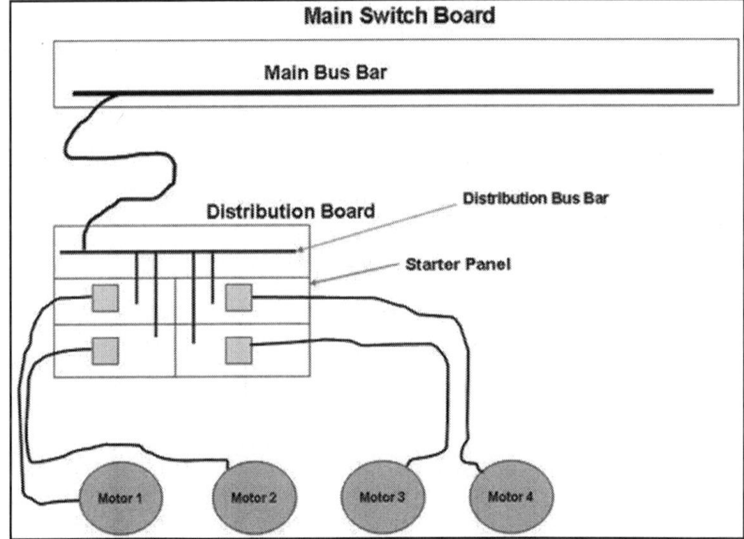

EMERGENCY SWITCH BOARD

As the name indicates, it supplies to equipment and machinery that are essential during any emergency like fire, flooding, etc. One important aspect about the emergency switch boards is that it is located above the load water line or the upper most continuous deck, where as the main generators and main switchboards are located below the weather deck or the load water line. Thus this ensures that the emergency supply is maintained when the vessel is flooded with water to the weather deck.

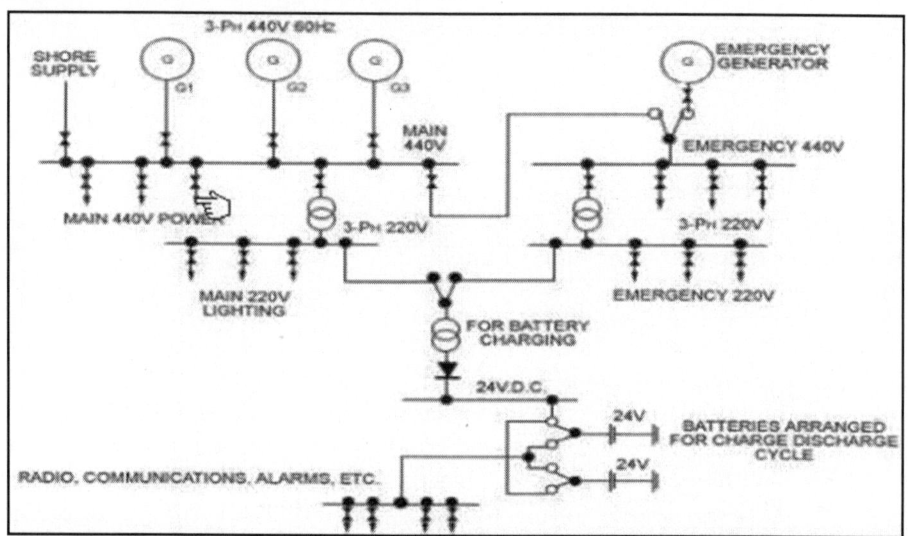

During normal operation, power is supplied from the main generators to the main switch board. The emergency switch board gets it supply from the main switch board.

When the main generators fail, the main switch board looses its supply and this causes the emergency generator to start automatically, which will supply the emergency switchboard that is connected to various ship board auxiliaries. Thus critical equipment and machinery (essential services) are supplied always to maintain the safety of the ship and personnel.

ROLE OF COMPRESSED AIR IN STARTING A MARINE DISEL ENGINE

Do you know how diesel engines are started in ships? Equivalent in size to a four-story building, the main propulsion engine is started with the help of compressed air at a pressure of 30 bar. Learn more about where a ship gets itscompressed air supply from.

We discussed diesel engine starting problems in ourprevious article and saw how various methodsare used to overcome this problem. We also saw how a marine engine is different due to its size and location, and that compressed air is the solution to starting the diesel engine. As you know that a ship is a mobile power plant or a moving mini-city. It has all facilities, sometimes better than what we find ashore.

These moving giants have a pre-designed and erectedcompressed air system, which facilitates many activitiesonboard a ship. There are mostly 4 to 8 and sometimes 10 air compressors found onboard. These aircompressors take suction from the engine roomatmosphere which is already under a slight positive pressure. These air compressorscompress the air in stages and fill up the huge air bottles, which acts as accumulator. They compress usually up to 30 bar and keep the air bottles filled up all the time. The number of air bottles and its volume(capacity) depends on the power (size) of the main propulsion engine. Air compressors are of varying capacity and used as per the requirement onboard.

ROLE OF COMPRESSED AIR ON A SHIP

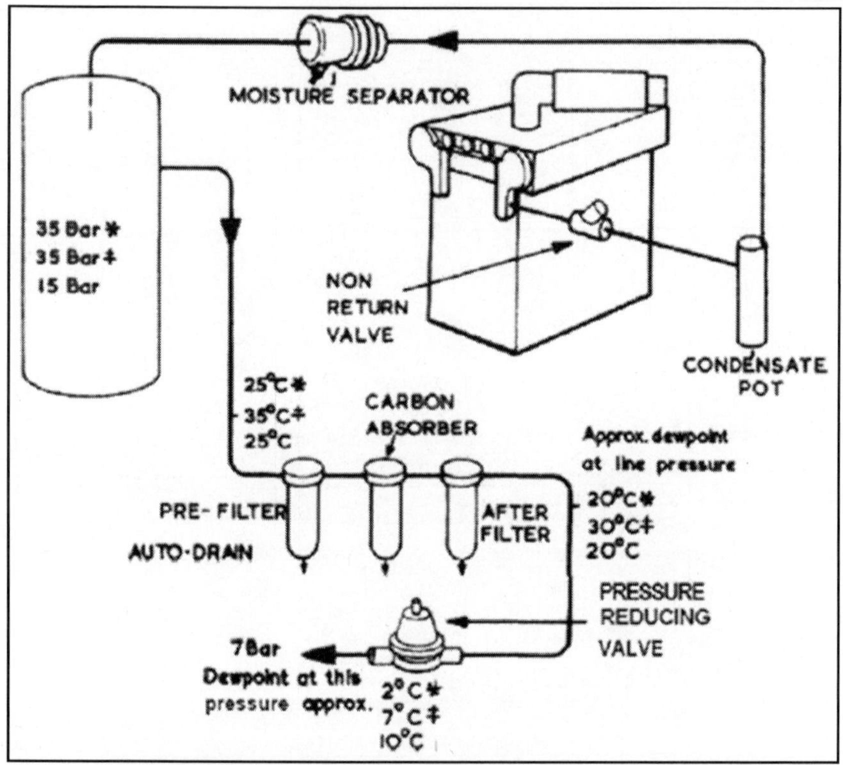

The role ofcompressed air onboard is of a very vast nature. Every ship will have a "DEAD-START" or "The FIRST START" arrangement. This is nothing but when the ship is totally "Dead" i.e, with out any power and no machineries running and no compressed air in the air bottle to start the generator engine(auxiliary diesel engine), then a provision is given for every ship to get the air bottle filled up to bring back the ship to safe, normal, working condition. This is provided either by a "Emergency Air Compressor" driven by a small diesel engine or electric motor getting its power supply from "Emergency Generator".

The various uses of compressed air onboard ship are listed below:

1. To start main propulsion engine
2. To start Auxiliary diesel engine(power generation)
3. To blow ship's whistle
4. For Engine Room general service and cleaning.
5. For the operation of pneumatic tools
6. For deck services, to carry out chipping.
7. For Automation & Instrumentation of various machineries,
8. For fresh & sea water hydrophores,

9. Fire alarms & operation of Quick closing Valves,
10. For Soot Blowing Exhaust Gas Economizer
11. And more!

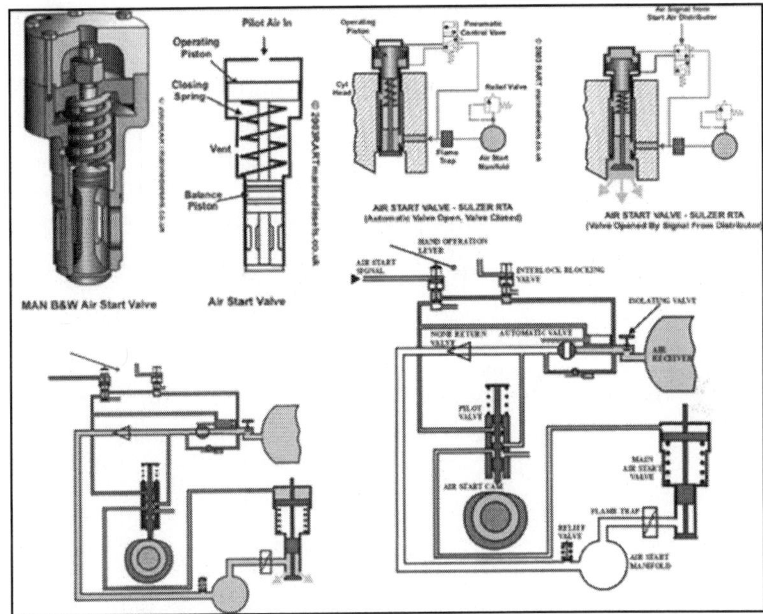

COMPRESSED AIR SYSTEM LAYOUT

The compressed air system onboard typically has a set of 4 to 6 compressors, out of which 3 will be main arecompressors, 2 service air compressors & one topping up air compressor. The Emergency air compressor is not to be counted as a normally used machinery.

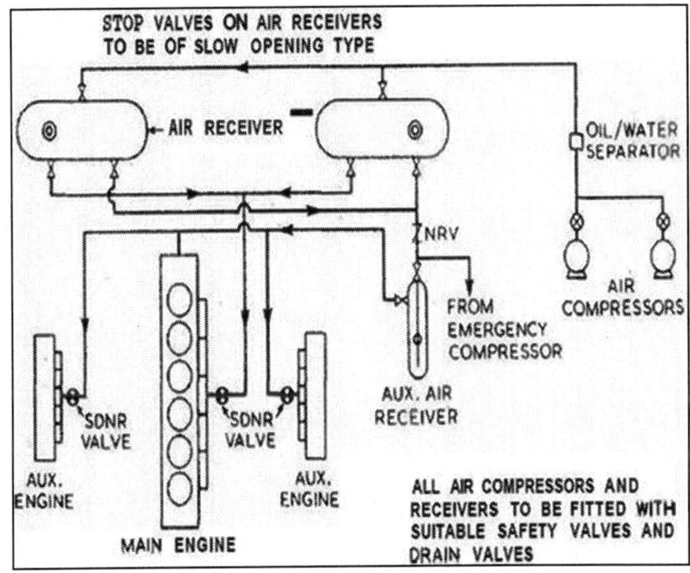

Main air compressors: These are used only when a ship enters or leaves a port, mostly during maneuvering only. They are of higher capacity of all. During maneuvering, these compressors when put in use, fills up the air bottle faster than other compressors, thus enabling the ship to berth herself in the port.

Topping-Up air compressors: These are comparatively of lesser capacity than the main air compressors. They are just used when the ship is sailing in the mid-seas where the consumption of air is very less (only for engine & deck services).

Service Air compressors: These are general service air compressors, which are used either for engine or deck service. They are of lesser capacity and their aredesigned for their pure strict oil free air quality. These compressors are used for pressing up the control air bottle and thus the control air of high purity is used for various automation purposes in the engine room and the pump room for oil tankers.

There may be set of main air bottles, service air bottle and a control air bottle for their respective purposes. The system also incorporates many pressure reducing valves and driers(de-humidifiers) to get rid of the moisture present in the compressed air.

STARTING OF AN AUXILIARY DIESEL ENGINE

The auxiliary diesel engine is mostly started with the help of compressed air,depending upon the size of the engine. Other means of starting includes Electric start(battery) and air motor(engaged in the flywheel). The most common method is the use of compressed air. The lay out for starting the auxiliary engine is given below.

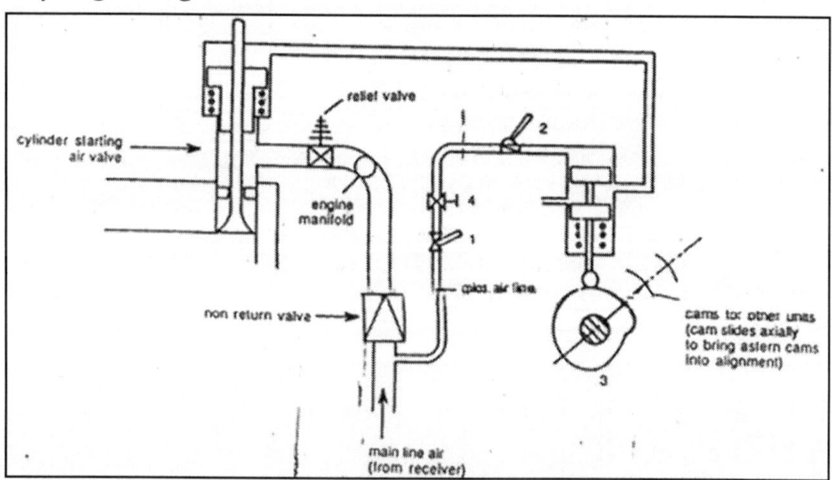

Before starting the auxiliary engine, the following safety checks must be carried out:

1. Turning gear disengaged(if available).

2. Lubricating oil sump level normal
3. Turbocharger oil level(both turbine & blower)side normal
4. Lubricating oil priming pump running.
5. Fuel oil/diesel oil booster pump running.
6. Lube oil, cooling fresh water, fuel oil pressure normal.
7. Rocker arm tank level normal.
8. All valves in compressed air line open to the engine.

Referring to the above starting diagram of an auxiliary engine, the" main air" from the main air bottle arrives at the air starting valve. There is a tapping from the main air starting line, "pilot air" going to the starting air distributor. When the engine rotates, the camshaft also rotates which in turn rotates "the starting air distributor cam". This cam isdesigned as per the firing order of the engine such that, the distributor rotates and lets the pilot air to the particular unit. The pilot air reaches on top of the air starting valve, opening it, in turn making the long awaited main air to let inside the combustion chamber. The main air which is at 30 bar, pushes the piston down making the crankshaft to rotate. This leads to continuous rotation of the crankshaft making the engine to achieve the minimum r.p.m at which firing of the injected fuel takes place. When the engine picks up on fuel, the air is cut off and drained. Thus the auxiliary diesel engine is started with the help of compressed air.

In the next article, we will take up the study of the various valves mentioned in the starting air systems namely master air starting valve, cylinder valves and so forth.

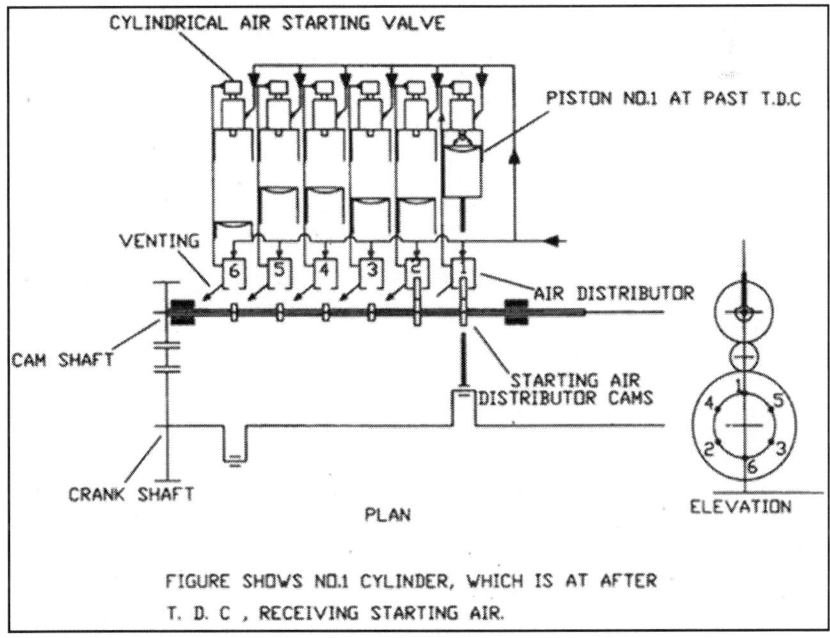

FIGURE SHOWS NO.1 CYLINDER, WHICH IS AT AFTER T. D. C , RECEIVING STARTING AIR.

ELECTRICAL SYSTEM USE AND MAINTENANCE

By the time a boat gets a few years on it, the amount of substandard and/or jury-rigged wiring on it can add up to substantial numbers of faults. Much to the consternation of the owner, when the insurance surveyor comes along, he ends up handing the owner a long, and frequently costly, list of things that need to be corrected. Even more seriously though, is what substandard wiring can do to your overall system, not the least of which are chronic battery drain, frequent system faults, or even causing electrolysis and fires. In the last month alone, I caught two boats that were about to go up in flames, and one that actually did catch fire during the course of a survey. One because the owner had installed air conditioning controls below a companionway hatch that got wet and shorted out, the other due to an improperly wired service outlet. Let's begin with the fact that a boat is not a house or a car, simply because the later two don't float in water, which is what makes a boat inherently different. Especially a boat that floats in sea water because sea water is a fair conductor of electricity. Therefore, there are lots of additional rules about the materials and methods of installation that you won't find on land based vehicles or structures. Because water (both inside and outside the boat) provides a very convenient ground path for electricity, we have to be a lot more careful how we do things.

MARINE ELECTRICIANS

There are no schools for boat electricians, only ship electrical engineers, except for the occasional vocational school. Most marine electricians are basically self-educated by means of attending seminars and the like. When hiring an electrician, you should not take their word for it because it says so in the ad, or your friend said Joe was really cheap. The least you can do is ask for a resume, since long experience is probably the best criteria.

MATERIALS & EQUIPMENT

It is important that only equipment and materials rated for marine application should be used. The American Boat and Yacht Council (ABYC) has very stringent standards for this, as does Underwriter's Laboratories. This means things like common wire and wiring devices, as well as electrical equipment, should not be used. That non-marine fan, airconditioner or battery charger, because the electrical apparatus is not properly constructed, insulated or grounded could either start a fire, or worse, result in an electrocution. Federal regulations require that all electrical devices used in the engine compartment or bilges be "ignition protected" meaning they won't create sparks that can cause a fire or explosion. One of the worst offenses we see on boats is the use of extension cords and other non marine wire, even such things as audio speaker wire and household lamp cord on high voltage systems. No, I'm not kidding. speaker wire carrying 125 volts. You should never use household wiring devices, such as multiple outlet expanders, or wire. Purchase electrical supplies only from a reputable marine supplier. Wire must have the proper temperature, water and oil resistant ratings and is hard to come by at retail outlets.

This service outlet was replaced by some one who didn't understand high voltage wiring and this is the result. The breaker was very old and the contacts fused closed and would no longer trip. This photo was taken during a survey while the wood behind the panel was burning. The owner showed no concern about it, thought it wasn't serious.

Here's a good reason why this ribbed plastic conduit should not be used on boats. Not only is it highly flammable, but it spreads the fire.

HIGH VOLTAGE SYSTEMS

No one who is not trained in marine high voltage systems should ever perform wiring on a boat. The potential for creating deadly system faults that are injurious to both people and the vessel itself is too risky. That includes something as apparently simple as wiring a new service outlet, for if you don't understand the principles of the system, you are likely to unknowingly create a problem. Just because you know how to do wiring around the house, doesn't mean you know what you're doing on a boat.

GROUNDS AND GROUNDING

One of the least understood aspects of a boats electrical system, and the most troublesome, is the proper method of grounding. That we often get questions of whether AC or DC electrical equipment should be grounded to the boat's bonding system is illustrative of this point. AC and DC grounding systems are two separate systems, for distinctly different reasons. If you don't understand these systems, you run the distinct risk of creating a disaster. Actually, there are four separate ground systems: DC ground, AC ground, AC grounding (or bond), and the vessel's bonding system. You can add to this lightning and HF radio grounds as well. Do you know the principles of each? Are you sufficiently confused to discourage you from doing your own wiring? I hope so. For unless you understand each thoroughly, you're headed for trouble.

The AC ground and grounding systems are "free floating," meaning that they do not ground on the vessel, but only to shore. The ground, or neutral, is a current carrying conductor, and is the source of many troubles because people do not regard it as such. The grounding, bond or green wire is the "safety" intended to channel current safely to ground in the event of a short circuit. Both of these circuits are capable of conducting current and can be the source of electrolysis when there are system faults with the dock or marina wiring. This is very easy to test for.

There is only one point where the DC side is grounded, and that is at the battery. It, too, is a "free floating" system in which nothing is ever grounded to any metallic part of the vessel, most especially not the bonding system. Just like a car sitting on rubber tires, completely insulated from earth potential, the battery itself provides the negative potential.

The bonding system, also green wire, has nothing to do with electrical systems. Underwater metals are simply wired together to equalize differences in potential of different kinds of metal. Nothing should ever be grounded to the bonding system. Unfortunately, some people don't understand this and use it to ground electrical equipment, occasionally with disastrous results.

BONDING SYSTEMS

Bonding simply means wiring all the boats underwater metals together.

This is done because of the galvanism caused by the different metals. By wiring them together, the differing potentials are equalized. Bonding does not solve problems of galvanism or electrolysis, but it does spread the the flow of current around over more metal, so that 1/4 volt or so won't cause any damage. Whereas if that 1/4 volt were going to one small seacock, it would probably eat it up in a hurry. In other words, bonding lessens the effect of small amounts of current. On the other hand, it also spreads it around to all underwater metals so that higher currents end up damaging everything.

Bonding systems use wire and ordinary crimped ring terminals. After a while these get wet and corroded. Electricity doesn't flow very well through corroded metal, so your bonding system after a while stops working. To maintain it, simply cut off the old terminals and install new ones. Do you have wires attached to sea cocks with hose clamps? Forget it. This is putting stainless and copper together, which are galvanically incompatible and it won't work.

YOUR BOTTOM PAINT

What does bottom paint have to do with electrical systems? Nowadays, with copper based paints, a lot. If, the next time your boat is hauled and you see large ugly burn patterns around all your underwater metals, you got a stray current problem. Copper-based bottom paints react severely to stray current, and serves as a great indicator. Sort of litmus paper for electrical problems.

Of course, the common wisdom is that the stray current "is from the marina." Or it's always the other guy's boat that is causing your problem. Don't bet on it. Most stray current problems are sourced on the boat in which they appear. Otherwise, everybody in the marina would have the same problem.

ELECTROLYSIS AND GALVANISM

Electrolysis is a word that is badly abused by boaters who don't really know what it means, so let me correct this right now. First, understand that all boats have an electrical potential.

That's because of all the different metals on the boat which, themselves have differing electrical potentials. This is exactly the same principle that makes a dry cell battery generate electricity. This electrical potential is called galvanism and is the reason why we put zincs on boats. Electrolysis is stray current escaping from the system and is most damaging. It is an abnormal condition. When this happens, it will eat up the zincs in no time, usually leaving that metal looking bright and shiny. Therefore: Shiny zincs = electrolysis. Dull eroded zincs = galvanism. I spent two years putting a meter on every boat that was hauled for survey. The average boat generates about 1/4 volt DC current and going as high as 1/3 volt without causing damage. But when it gets up to 1/2 volt, you got a problem. Zincs will erode rapidly and underwater metals begin to be affected.

SHORE POWER CORDS

The single largest cause of problems with shore power systems results from failure to maintain the connectors on both the cord and the boat connectors. These devices are exposed to water and over time suffer from corrosion and general wear. High resistance caused by corroded, bent or worn connectors results in high resistance which causes overheating, which further amplifies the power drop. This not only creates conditions for a potential fire, but causes electrical equipment to work harder, resulting in reduced life span of equipment. IT PAYS TO MAINTAIN SHORE POWER CONNECTIONS.

You can perform a very simple check just by placing your hand on the shore cord near the connection to determine if it is heating up. Obviously, this should be done while you have a lot of equipment turned on. If it's anything but slightly warm, not more than 110 degrees, suspect a problem. Shore power connectors should be dismantled at least once per year, cleaned and repaired as necessary. Most of these connectors have replaceable parts. If you drop your shorepower connector in the water, you must take it apart, clean and dry it. Otherwise, expect it to burn up. We recommend that you buy only the highest quality power cords, as these will last longer and have the advantage of replaceable connector parts. Cheap connectors usually can't be taken apart. We also advise against ever using the three-pronger household type adapters as this type of connector is highly unreliable and prone to causing system faults and fires. Only the twist-lock type connector is suitable. One more thing: If you are not turning off the dock breaker before disconnecting the power cord, start doing it now. Not only do you risk getting electrocuted, but disconnecting an energized connector damages the contacts. Also consider what happens if you drop the energized cord in the drink!

POLARITY

This is an issue only with 125 VAC systems since 250 VAC systems will not function with wrong polarity. Since you have three terminals on a shore connection, wrong polarity can mean that any of these wires are in the wrong position. Not only should you pay attention to the polarity indicator on your boat, we recommend that you keep a plug in polarity indicator aboard and use it every time you hook up to shore power at a different location.

Reverse polarity is not only an electrocution hazard, but can also damage electrical equipment. It is most often found with the three prong spade connectors (household type), but occasionally twist lock connectors as well, particularly in marinas with dilapidated equipment. Never trust the power supply at strange docks; but always check the polarity. When hooking up to strange docks, always check your volt meters to make sure you have adequate voltage. Low voltage is very damaging to electrical equipment. Turn on the stove or water heater and watch what happens to the meter.

MAIN CIRCUIT PROTECTION

Many people think that the circuit breakers on the dock protect their boat. They do not; they only protect the dock wiring. Your main circuit breaker protects your boat's systems. But what about that section of wiring and connectors between your main panel and the dock breaker? Well, the fact is that it is unprotected. which is why so many fires occur. Check out all the top end boats and you will find that they have circuit protection located directly at the shore connectors. Which is why we recommend that you should too. Having slow blow cartridge fuses installed directly at the connectors can go a long way toward preventing fires and burned up shore cords, particularly if you are a traveler and frequently rely on uncertain power supplies. Circuit breakers should NEVER be installed on the exterior of the boat. Only gasketed, water proof cartridge holders should be used.

CIRCUIT BREAKERS

Circuit breakers wear out, and when they do they work less well, or not at all. If you are using circuit breakers as ON/OFF switches, you are helping them wear out that much faster. It also damages breakers when you shut off equipment via the breaker. This causes arcing at the contact points which damages the points. When connecting and disconnecting shore power, you should always turn OFF equipment at the appropriate switch on the equipment. Then shut the main breaker off. Do not ever simply throw the main breaker off to shut down equipment that is operating. The circuit breaker arcs and damages it.

Also be aware that any equipment run by a motor, such as air conditioning and refrigeration equipment, start up with an initially much higher amperage than the normal running amperage. An air conditioner that runs at 14 amps may have a start up amperage of 20 amps, so that if you just go and turn all the equipment on at once, it overloads the system. Then the circuit breaker gets hot and won't stay engaged until it cools down. Ergo, start up heavy equipment one item at a time, allowing it time to cycle into its normal operating voltage before turning something else on. For example, don't turn the AC, refrigerator and icemaker all on at once and not expect the breaker to pop.

CHRONIC BREAKER POPPING

It's human nature that when a problem appears, we wish it to go away. Breakers that pop frequently are signaling that there is a problem, which could either be the breaker, or something in the circuit. Yet most people will keep on attempting to make the breaker engage. This can be dangerous because you may cause the contact points of the breaker to fuse together from arcing, in which case it will never trip again. The above photo shows what can happen when you do this. DO NOT ATTEMPT TO ENGAGE A BREAKER THAT IS OVERHEATED BY FORCING IT. You must allow it to cool down.

If you are experiencing chronic problems with circuit breakers popping, first check how much current draw is involved. A single 30 amp circuit is not much when you're running things like air conditioners, water heaters and battery chargers. One very simple way to check whether you're dealing with an overload problem is to add up the amperage draw of each piece of equipment. List both the start up and run amperages. You will usually find the amperage given right on the equipment label. By making a list of the total power demand, you'll get a good idea of what you can and cannot operate simultaneously, particularly when starting the equipment. If you have an ammeter on your panel, check it against the amperage tally you made. Ideally, you should try to hold power consumption at 80% or less than the line rating.

Check the breaker by allowing it one hour (or whatever it takes) to cool down. Turn the equipment off and, after it is cool, reengage the breaker. Now turn the equipment back on. Place your finger on the front of the breaker and note its temperature. If it does not heat back up again, then the problem was probably a start-up overload. If the temperature rises again, there is a fault in the circuit or the breaker. (Note: when the breaker contact points become eroded, the breaker itself can overheat).

GFCI'S

Ground fault current interrupter service outlets are required to be installed in wet locations such as the galley or head. In reality, there's little chance of being electrocuted inside a boat because you are not grounded within the boat. A greater risk is from service outlets being located in places that get wet, such as below leaking windows, hatches or close to doors. Three pronged plugs are prone to shorting across the terminals when wet, so having all your service outlets changed to GFCI's is a good idea. Use only the highest quality devices from a reliable manufacturer like GE, and not the el cheapos from the local discount store. They aren't worth having. Service outlets located anywhere on the exterior of the vessel are an invitation to trouble for reasons that should be obvious.

DC SYSTEMS

The same advice about jury-rigging wiring applies to DC systems as well AC systems. While you're not going to create an electrocution hazard, it is very easy to take a faultless system and create faults in it. A typical problem starts like this: The owner wants to add a new piece of equipment, but the electric panel is way over there, and the place he wants to install the equipment is way over here. Besides, there are no extra breakers in the panel, and no space in the panel to add another one. To make matters worse, the panel is located in such a way that he couldn't string new wires into it even if he wanted to. So what he does is to find a place where he can tap off an existing circuit, and maybe adds an in-line fuse, stringing wires all over the place in the process. Or maybe he is replacing a piece of equipment that has a faulty circuit, but instead of trying to locate the fault, he just clips off the old wires and strings new ones. This happens a lot, and by the time the boat has a few years on it, it's got cut wires all over the place, many of which are still hot!

In many cases, he will just go and take new leads off the batteries, bypassing the panel altogether. Now when he goes to turn off the main power supply, all that new stuff added remains energized. In addition to which in-line fuses have also been added all over the place, so when something craps out he's got to go tearing through the boat to find that hidden fuse.

If you have electrical problems and your system looks something like this, then you needn't look much farther for the source of the problem<

Not every electrical system is going to be this neat, but this is the way it should be.

The worst of the problems with DC system add-ons comes with improperly installed wiring and the use of wire splices of all sorts. Typical of these are the use of electrical taped connections which, when the tape gets warm (as in the engine room) the tape glue gets soft and the tape falls off. Or the use of wire nuts or crimped butt connectors in locations that get wet. Wire nuts (those twist-on cones) are not approved for marine use. When connections get wet, the wire corrodes, creates high resistance, usually resulting in equipment damage or failure for reasons which the owner will never discover. He'll think just that damned lousy piece of equipment crapped out, when in fact the lousy wiring job is the culprit.

PRINCIPLES OF WIRING

I would venture to say that half the electrical problems on boats result from improperly installed wiring. After the boat is built, there's no convenient way to route new wiring. But we need to understand that systems on boats are subject to high G-forces due to pounding, rolling and vibration. Connections get stressed and wires rub and chafe against abrasive or sharp objects. It doesn't take much damage to wire insulation before you have a condition where stray current may develop. And the chance of finding a little bit of damage on one wire is about nil.

1. Must be routed in a suitable, dry area and be well secured. Should not be laying in bilge or in areas that get wet.
2. Must not be routed with pipes or hoses of any kind, and not be in contact with fuel tanks or fuel lines.
3. Splicing circuits should be avoided. If splicing is necessary, it should employ a proper terminal block, and not butt connectors. Every splice in a circuit creates additional resistance, and the potential for the connection to come apart. Taped connections and wire nuts should not be used.
4. Wiring must be firmly secured and in locations where it won't get damaged. Should not be dangling or strung across open spaces. Use only plastic, not metal, clips to secure the wiring.
5. Must have chafing protection or conduit at vibration points around machinery.
6. Must not be in contact with, or proximity to machinery exhaust systems.
7. Wiring should be neat. A boat full of tangled wiring demonstrates unprofessionalism and the inability to fix something that goes wrong. An electrician can't trace a plate of spaghetti, and when something does go wrong, the cost of fixing it goes way up.

CIRCUIT PROTECTION

Adding in-line fuses to a newly installed piece of equipment is a terrible way to add circuit protection. First of all, these devices trap water and corrode internally. Secondly, you end up with two more splices in a wire circuit that shouldn't have any. Third, you usually forget where they're located, and if you've got ten of them on your boat . . . well, you get the picture. Jury-rigged systems are just that; a temporary, unreliable system. A boat full of in-line fuses is a boat full of short cuts and amateur installations.

BATTERIES

Batteries are a constant source of aggravation to many boat owners, almost always for reasons that are preventable. These are: low quality batteries, poor or non existent maintenance, and improper installation and wiring. As a general rule, batteries perform consistent with the price you pay for them. Good batteries are expensive, and shopping for price will only lead to momentary satisfaction. Cheap batteries have thin plates and poorly insulated casings and therefore cannot give long service.

This illustrates why crimp-on ring terminals, wing nuts and other substandard high amperage cable connectors should never be used. High resistance has melted the cable insulation and burned up a $900 starting motor because of high resistance. Secondly, batteries have to be installed correctly. That means in a clean, dry location that can be reached. If you can't reach them, if you have to kill yourself to get at them, then you will not maintain them. If installed in an inaccessible location, you should consider having them moved to a better location.

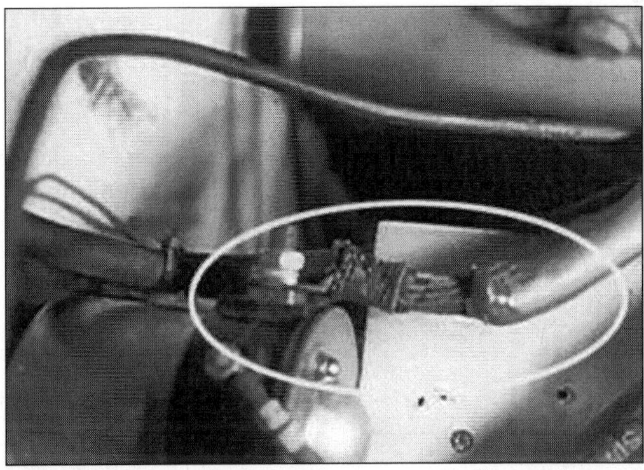

Place a fully charged battery on the concrete floor of your garage. Then come back two weeks later and check the charge. That battery will have completely discharged, and it will have done so right through the plastic casing. Now you understand my point about proper installation and dryness. On many small boats, I usually find the batteries sitting in uncovered plastic boxes that are full of water. Or they're sitting in bilge water or on wet decks. If that's the case, you needn't look any farther for at least part of your problem. If you want your batteries to be reliable, they must be kept clean and dry. That includes the top surface, particularly between the terminals.

PROPER LUGS

It won't do to make your cable connections with threaded studs and wing nuts. These afford inadequate contact surface that can cause high resistance and is one of the major causes of engine starting motors burning out. Your starting motor cables should be attached only with swaged lead lugs, not the ring terminal kind smashed with a hammer to make the connection. Small boats are usually the worst offenders in this regard.

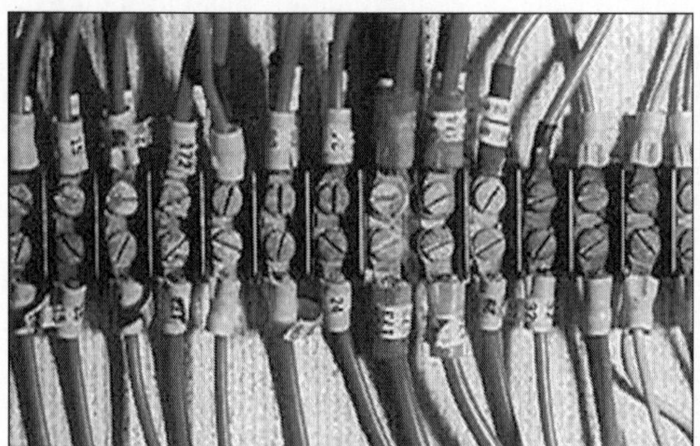

Corrosion takes its toll even on the interior of the boat due to leaks, salt air and high humidity. When tested, most of these connections had high resistance. This is the proper method to splice wires.

VENTILATION

Batteries develop heat when charging, as well as hydrogen gas. For that reason, the need to be in a well ventilated area. Gel cells are particularly vulnerable to overheating damage. Putting gel cells in covered, plastic boxes has proved to be a problem, and for this reason they are falling out of favor. Sometimes improved technology isn't an improvement after all.

ELECTRICAL DEVICES EXPOSED TO WEATHER

Why it is that there so many builders that install instruments, panels and switches in locations that are going to get wet is something I'll never understand. Oh, I know, you look at those switches and think that they're water proof. Well, you just go look at the back side of the panel and see if you still think so. See if you don't see a lot of corrosion back there. What happens when the back side gets wet? Well, water being a conductor means that these devices will short small amounts of current across the terminals, or to any available ground. This is one of the reasons why you have so many engine instrument failures, and boats have so much of a hardware corrosion problem, and why they have stray current problems, never mind equipment failures. Electrical equipment exposed to weather that is not absolutely water proof is just asking for trouble.

LOCATING INTERNAL EQUIPMENT

Just because it's inside the boat doesn't mean that electrical equipment will stay dry. Boats leak, and stuff located under those leaks are going to get wet. That means that you have to pay attention to where you put stuff. Many people mount various types of pumps low in the bilge, assuming that the bilge water is never going to rise. Believe me, the bottom of the boat is the last place you ever want to mount something. Sooner or later your bilge pump will fail, and when it does that expensive equipment is going to get ruined.

The other place you never want to mount anything is under an open cockpit deck. For reasons I won't get into here, this is a terribly wet environment, which is why boats with generators located here end up replacing generators a lot. If you have exposed panels you need to keep them covered and dry. The vast majority of small boats that I see don't even have panel covers, yet alone having covers that are used. Having a small cover made up is a small price to pay to avoid serious damage and other problems.

Looking at this ball of wiring hanging down, is it any wonder that a lot of electrical equipment on this boat didn't work? Instrument panel of this open boat was not kept covered. Water pouring through the panel had even filled up the battery boxes with water.

Close up of wire bundle at left. These supposedly water proof wire connectors, on closer examination, are found to be full of water.

ENGINE WIRING

Due to vibration and high temperatures, damaged wiring on and around engines is one of the most common causes of stray current damage, i.e. true electrolysis. It is extremely important to consider the routing of the wiring so that it is not in contact with hot manifolds, or vibrating on sharp edges or rough surfaces. Engine wiring should be inspected periodically for signs of damage. Avoid using the kind of highly flammable plastic conduit shown above. If you are painting the engine, do not paint the wires; the solvents in the paint causes the insulation to become brittle and crack.

THE EFFECTS OF TIME

Electrical systems don't last forever. Over time boat systems degrade, particular as respects to the effects of corrosion. We recommend that a boat should have an electrical system survey once every 5 years. This does not take long and does not cost much, and can save you thousands of dollars in unwelcome headaches. In most cases, the survey will simply reveal the effects of corrosion which needs to be cleaned up, along with few possible minor repairs. The advantage is in not allowing problems to accumulate to the point that it becomes a big, costly job.

A WORD ABOUT BUYING OLDER BOATS

Everything I've said so far comes to bear on what you are likely to obtain when you buy an older boat. One of the things that we pay close attention to in performing a survey on a used boat is how much jury-rigging and alterations there are in the electrical system. Sometimes there's a lot, and if the new owner isn't willing to foot the bill to make corrections, more often than not he's going to be living with a lot of electrical headaches. In most cases, he's not even aware of why things are constantly breaking down; he just knows that things unexpectedly keep crapping out. Usually he blames the pump or motor, or whatever gizmo that fails, when actually faulty wiring is the problem. Beware that on boats with bastardized electrical systems, the cost of straightening it out and cleaning it up can be pretty high.

ADDING NEW EQUIPMENT

Most boats, particularly smaller ones, are unfortunately not designed with the idea of adding equipment, even though nearly everyone does. But the primary reason why systems gets so messed up results from the attempt to force something in where there is no provision for it.

The best way to deal with this is to simply add a new panel, whether AC or DC. Doing this is neither very difficult nor expensive, especially considering the cost to repair all the problems you're likely to create when you try to cram something onto a system not designed to handle it.

A six breaker DC panel will cost less than $200 and is a whole lot better than scattering inline fuses all over the place, and with installation will run less than $500. Plus, the new panel can be added at any location where it will be convenient.

Adding a second shore power circuit is especially useful for those boats with single 30 amp shorelines where the addition of new equipment will tax the system. It's not very convenient to have to turn the water heater off to turn the air conditioning on, or constantly be managing equipment as breakers keep popping off. You have two options here, the first of which is to increase your shoreline to 50 amps. To do that you have to change the main breaker, power receptacle and the shore cord, which is a lot bigger and heavier. The other option is to add a second 30 amp circuit, which means adding a new panel, receptacle and shore cord, which costs a bit more.

Adding a second circuit has the advantage of separating the air conditioning onto a separate line, which is the way all good systems are designed. That's because of the high power demand, combined with frequent dock power faults, makes it desirable to separate it from the main service.

UNDERSTANDING MARINE ELECTRICAL SYSTEMS

The electrical system on your boat is the power behind the power. Without it, the outboards won't start, the pumps won't pump, the lights won't illuminate and the navigational electronics show only black screen. Many boat owners don't have a basic knowledge of the electrical system on their vessels and that lack of knowledge can result in lost days on the water or worse. Today's outboard powered boats have sophisticated power distribution systems that incorporate multiple batteries, isolators, chargers, breakers, switches and fuses - you should understand the function of each.

Power systems vary with the size of the boat, the number of engines and the power demands of the systems on board. Vessels with multiple outboard engines and a larger compliment of electrically-powered accessories require a more robust and complex system than a single-outboard skiff. Keep in mind that not all of the components covered in this issue are found on all outboard boats.

Joe Vizzosi recently took delivery of a new Yamaha-powered 39-foot Contender® with three F350 outboards. While he only had the boat for a week when he welcomed us aboard, he was already able to walk us through the components of the electrical system, from the batteries to the breakers at the helm switch panel, allowing us to shoot pictures along the way. This boat has a far more complex electrical system than smaller outboard-powered boats with less engines and batteries. Vizzosi's understanding of the systems onboard was impressive.

"It's critical to know where everything is located and its function in the system," he said. "While electrical systems on newer boats are pretty foolproof and trouble-free, you never know when you'll run into a problem with the power supply to your navigational electronics or running lights. If that happens offshore, you have to be able to fix it."

Here's a brief rundown of the major components commonly found in modern outboard-powered boats.

CRANKING BATTERIES

Cranking batteries are dedicated only to starting your outboard engines. Once they do their job, they're immediately recharged by the engine alternator. They are designed to provide a burst of amperage to the starter motor and, therefore, must be capable of providing the cold cranking amperage (CCA) required by the engine manufacturer to accomplish the job.

CCA is a rating that defines a battery's ability to start an engine at 0°F (-17.8°C), and can range from 250 CCA for low horsepower electric start outboards to 750 CCA for high horsepower V6 and V8 outboards. New boats come equipped with appropriately sized cranking batteries, but be sure to check your outboard owner's manual when the time comes to replace old batteries.

STORAGE BATTERIES

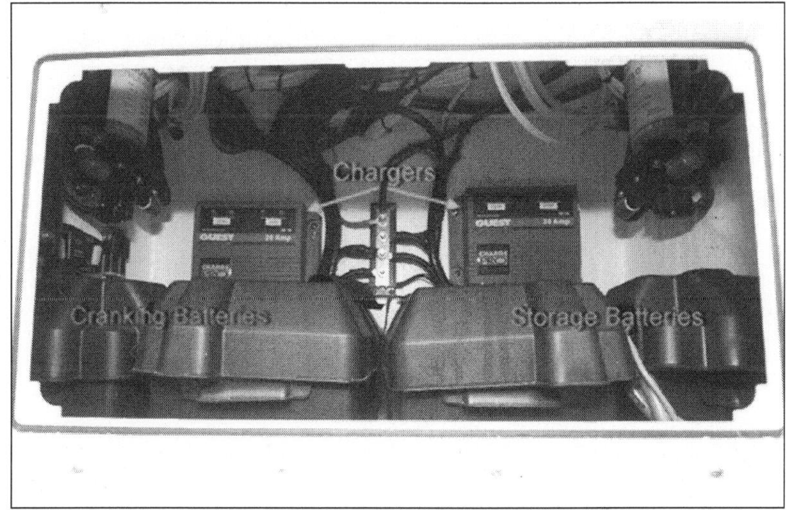

Also called house batteries, storage batteries provide the power to run all the boat's electrical accessories like running lights, bilge and bait well pumps, navigational and communications electronics, entertainment systems, anchor windlass, bow thruster, power steering, and air conditioning. Storage batteries also power refrigeration on larger boats or electric trolling motors on bass and walleye boats. This system can range from a single battery on smaller boats to a bank of batteries on boats that require more amperage, and they are typically deep-cycle type.

Deep cycle batteries are designed to provide constant power over long periods of time and, unlike cranking batteries, are capable of withstanding extreme discharges and recharges without damage.

BATTERY ISOLATOR

A battery isolator is a single-direction pathway used for directing the power from the engine's alternator to two or more batteries. For example, a vessel with twin outboards charging two cranking batteries and a house battery has the current from the alternators pass through the isolator, which distributes it to the batteries. Battery isolators also prevent competing batteries from discharging from one to the other during operation.

BATTERY SWITCH

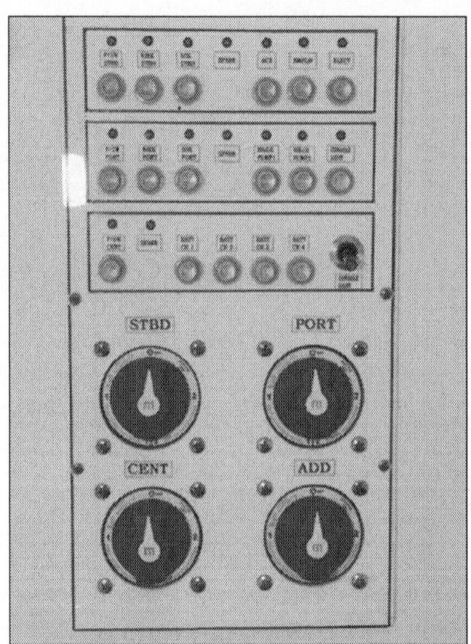

A battery switch, or switches in the case of multiple outboard applications, is used to engage or disengage the various batteries on the boat. When the boat is not in use, they are turned off. There are switches designed for single or twin battery applications. For example, a single-outboard boat with two batteries can be operated from a dual battery switch, which can engage them individually (battery 1 or battery 2) or in unison (both). Boats with more than two batteries typically employ a switch for each battery in the system with just an on-off configuration.

SMART CHARGERS

Smart chargers are more popular than ever for maintaining batteries at maximum capacity when a vessel is not in use. They are available in configurations capable of handling any number of batteries. Smart chargers are installed on the boat with a receptacle for plugging in a shore power cable for connecting them to a land-based 110-volt power supply. They distribute power on an as-needed basis to all the batteries on the boat. They are called "smart chargers" because they automatically detect when a battery falls below full charge, and then they send just the right amount of current to bring it back to full charge without overcharging.

CIRCUIT BREAKERS

The circuit breakers on a boat do the same job as the circuit breakers in your house or apartment. They are an automatic emergency power cut-off in case of a power surge, providing protection for the system and electrical components. Circuit breakers can be found in various locations, not just at the switch panel at the helm. You should always know where all the breakers on your boat are located in case one trips. This will help you to troubleshoot a problem and reset them to restore power to a circuit.

FUSES

The fuses are the last line of defense for accessories like your navigational and communications electronics that are not on individual breakers. A fuse panel is usually found in the vicinity of these items and can be traced by following the power cables from the electronics back to the panel. Most newer boats use automotive-type fuses, but many older vessels are equipped with glass fuses. You should make it a point to carry replacements for every fuse on the boat in case of a failure.

The list above details the common components that make up the power grid on your boat. You should take the time to familiarize yourself with the entire system, from the power source to the batteries to the distribution panels. Knowing and understanding your electrical system is as important as making sure there is gas in the tank before you head out for a day of boating.

3

Electrical Power

SHIP POWER PLANTS - BRIEF OVERVIEW

All systems and mechanisms of ship power plants (SPP) can for convenience be divided as follows:

- Main plant used for ship propulsion;
- Auxiliary plant used for energy supply at moorage and for start-up of the main plant;
- Electrical used for electric energy supply;
- General purpose systems and mechanisms.

Standard SPP transform heat energy into mechanical work or electric energy. Main SPP consists of the main engine, main gear, shafting and propulsive agent. All these elements as a whole are machine-propulsive complex. Continuous work of the main SPP is provided by feeding systems of fuel, oil for mechanisms oiling, cooling systems, air-gas system, process air supply system.

Main SPP are group by various features:

- By type of the main engines of SPP:
  - Installations with internal combustion engine, in general, diesels ones;
  - Steam engines;
  - Gas and steam turbines;
  - Nuclear power plants (NPP);
  - And mixed plants, i.e. comprised of several plants of various types;
- By type of propulsion shafts – one, two or multi-shaft (most often are installed at large war ships);
- By number of engines: one, two and multi-engine (two-engine are used at large vessels, and multi-engine – at vessels with electrical main gears);
- By type of the main gear – with direct, mechanical, electrical, combined and complex gear;

- By type of the propulsion device – propellers, hydraulic jets, airscrews, and vane propellers.

There are other classification features.

Auxiliary SPP consists of ship electric power plant (SEPP) and auxiliary boiler plant (ABP) producing low pressure steam. As a SEPP, internal combustion engines, steam or gas turbines can be used.

SHIP DIESEL PLANTS

Today, about 90% of the world civil fleet is equipped with diesel SPP (56% in Russia). The first diesel SPP ship was built in Russia, only few years after invention of this engine. Time tells that owing comparatively small dimensions and high productivity, such plants meet all daily wants of the fleet in full. And it is necessary to make a special emphasis on diesel-electrical power plants (DEPP). Though they are less saving comparing to diesel ones, they are exemplar when frequent change of propeller rotation mode is necessary. These SPP feature in torque moment equal for fore and back run, and at small speed it is bigger comparing to diesel SPP with mechanical gear.

STIRLING ENGINE

Today Stirling engines are being enhanced which are more saving and ecology-friendly comparing to diesel engines. These engines use external heat supply (of every type – starting from heat of end product of fuel combustion and ending with solar energy). Fluids and gases can be used as a working medium. The main limitation of such engine is need to keep up ultra-high pressure as to maintain parameters comparable with characteristics of internal combustion engine, and big square of heat-exchange elements. The power plant of such type is noiseless, provides low vibrations, has long life due to simple construction and works at minimum temperature gradient. At the moment, the submarine fleet of Sweden is equipped with such engines. The Russian submarine fleet also performs research in this direction.

GAS TURBINE

Gas turbine (GT) consists of a compressor, combustor and a turbine itself. GT comprises advantages of the diesel engine and steam turbine. As a diesel, GT does not need boilers to produce steam, as a steam turbine, it has no reciprocal elements. With its small weight, GT produces more energy than diesel PP, has short setting-up procedures and high acceleration. Its main disadvantage is diseconomy which can be mastered by means of raising temperature of exhaust gases and usage of exhaust heat in electric energy production. For example, effective thermoelectric semiconductor transducer can help with it. Its operating principle is based on generation of electromotive force while samaric sulphide (SmS) semiconductor is being uniformly heated without external temperature gradients. Coefficient of efficiency of the working sample is over 40%, so commercial production is possible.

STEAM TURBINE

Steam turbine is an engine where the supplied steam heat is converted into kinetic energy used for mechanical work. Today, steam-turbine plants as main SPP are used only at large war ships and high speed and large container ships with main engine over 40 000 hp. As part of complex PP, ST are used at nuclear ships.

NUCLEAR POWER PLANTS

Ship nuclear power plant (SNPP) consists of a nuclear reactor, steam-generator and steam turbine. Controlled fission reaction aimed at production of enormous amount of heat is led in the reactor. Via the cooling circle, the heat is supplied to the steam-generator; the produced steam is supplied to the turbine where gained kinetic energy is converted into mechanical work. Undoubtedly, SNPP differs in high capacity and possibility of long autonomous work. Its disadvantages are high cost of construction and servicing.

The first civil ship with SNPP (Lenin icebreaker) was built in the USSR. Being the leader in the field of nuclear ship building, Russia continues enhancement of SNPP.

Thus, nuclear icebreaker ËÊ-60ß which is in construction at the moment will be equipped with new Ritm 200 reactor plant which differs in gravity circulation raised up to 60% requiring application of less power pumps. The capacity factor of this plant is 0.75, safety factor is also elevated. Core refueling is going to be held once per 7 years, while modern icebreakers this period equals to 2.5-3 years.

DEVELOPMENT PROSPECTS

Modern nuclear plants with water cooling are futureless comparing to plants with liquid metal cooled reactors. Such scheme was at first used at Sea Wolf submarine of the US navy; as a primary circuit cooler they used a mix of fusible metals potassium and sodium. In our country lead-bismuth alloy is used as a cooler.

This scheme can be used with paraffin-like organic materials (diphenyl and triphenyl), but there are problems of corrosion or resinous material formation.

Another problem being solved by Russian engineers is direct conversion of heat energy into electric by means of thermoelectrical (TEG), magnetohydrodynamic (MUDG) or thermoemissive (TEG) generators. Environmental degradation encourages usage of low-waste, ecology-friendly SPP. At the Far East accumulator or, at least, mixed diesel-accumulator plants are proposed for usage at coastal vessels (i.e. passenger). In addition, accumulators can be charged at night via port power supply networks which will help to solve the problem of energy redistribution.

MARINE ENGINES AND PROPULSION SYSTEMS – BRIEF OVERVIEW

Marine Engines and Propulsion Systems – Brief Overview

The diesel engine is the most popular type of the marine engine.

Traditionally, marine diesel engines are divided into low-speed (piston speed up to 7 m/s), medium-speed (7-10 m/s), and high-speed (over 10 m/s). Besides, they are divided into slow (up to 350 rpm), medium (350-750 rpm), and high running (over 750 rpm) engines. They can also be classified by fuel injection, blow-down type, etc.

Marine ships are equipped with the following engines:

- Simple-acting engines with cross, loop, uniflow valve scavenging;
- Engine with uniflow scavenging and diverging pistons;
- Double acting diesel engines.

Since marine diesel engines are to ensure not only forward but even back run of propellers both directly and via reduction gearing, they differ in some auxiliary components from surface engines.

DEVELOPMENT PROSPECTS

Further modernization of ship engines is supposed to be made by new various solutions. Thus, two-stage boost allows reducing of engine mass-to-power ratio 1.4-1.5 times while effective cooling of cylinder head and boss will ensure optimum combustion space thermal environment. Application of long-stroke slow-running diesel engines with 22% piston upstroke allows significant decreasing of crank axle rate speed thus saving 5-8% of fuel. The deflation valve hydraulic actuating system and air damper will replace springs. Slow running engines will be retrofitted by enhancement of air-fuel mixing, fuel feed, and turbo-charging application. Some improvement feasibilities of diesel-electric power plants for submarine fleet are under consideration. The first is related with application of high-running (1200-1800 rpm) engines with total aggregate capacity of 700-1500 kW as alternating current generators, upgrading of electric machine efficiency and capacity, refinement of traditional batteries and development of new ones. Works in the field of application of high temperature superconductors contribute to creation of efficient non-nuclear power plants in future. The second improvement feasibility is related to application of anaerobic SPP: Stirling engines, direct conversion of chemical power to electricity without any interim steps, and enclosed diesel power plants.

TURBOMOTORS

Propulsive systems equipped with gas or steam turbines excel with major moving force. The capacity of such system in tow can be higher than in lost motion. In addition, turbines feature in torque self-adjusting thus excluding the possibility of overload.

Gas turbines (GT) disadvantage mainly in low efficiency (about 30%), high fuel consumption, sensibility for assemblage and fuel preparation as well as impossibility of reversing; which is the reason of installation of additional turbine aboard GT ships. All of this retards introduction of turbines in the civil shipbuilding. However today in addition to navy ships, such PP are installed at megatankers, container carriers, aerostatic crafts, and hydrofoil vessels, that is at vessels requiring high capacity. Besides, turbines of many ships are used as boost installations (afterburning plants).

PROPULSIVE SYSTEMS

Propulsion devices are designed to assure ship motion by conversion of some kind of energy. They can be active and reactive. Sails are active propulsion devices, others are reactive.

Today, the following propulsion devices are in active usage:

- Vane ones – pulling oar, paddle wheel, propeller screw, cycloidal propulsion unit;
- Water-flow – thrust and water-jet motor.

Today, the most ancient propulsive devices are used at sport vessels and boats – the oar – and at sport cruisers and pleasure yachts – the sail, while the paddle-wheel is substituted by the propeller screw. The propeller screw is the most popular of propulsion devices, and is referred to hydro-reactive ones. It consists of a hub with 2-6 blades equally spaced in a circumferential direction.

By design, propeller screws can be:

- Fixed pitch propellers (FPP), and
- Adjustable pitch propellers (APP).

The propeller screw is used at various ships – from transport to submarines. Their disadvantage is need for increasing of screw dimension and weight in order to up-shift of impulse to the vessel. Speed increase entails cavitating which threatens with blade destruction or loss of screw efficiency. In the middle of the past century, the way out was offered – operation in supercavitation mode. The efficiency of special supercavitating screws is 15-20% higher than that of standard screws. Such screws can effectively operate at 3000 rpm, so they are used at direct-acting gearless "engine-propulsion device" systems.

The cycloidal propulsion unit is also hydro-reactive.

It is a flush-bilge disk with propeller-type blades.

The disk circles the vertical axis while blades move rotationally oscillatory. Such propulsion unit advantages in high maneuvering ability of the vessel. Two symmetrical cycloidal propulsion units allow changing speed and direction of movement up to and including full stop without reversal of the main ship power plant.

Their disadvantage is low efficiency, structural complexity, limited transmission capacity and inefficient operation in rocking conditions, frequent blade crash and general expensiveness. So such propulsion units are installed at highly-maneuverable ships (ferries, bibcocks, craft tugs). Cycloidal propulsion units are also used as auxiliary thrusting propellers at large sea-going vessels. Water-jet propulsor. Water-jet propulsor consists of a pump or screw, shaft, water pipe (water-jet pipe), counter-propeller (outlet straightener), and reverse-steering arrangement. Water ejected via nozzle provides movement of the ship.

Since absence of extensions assures good ship manoueuvrability in shallow water areas, water-jet propulsors are suited to shallow-draft ships (river-going vessels, for example). Currently, water-jet propulsors for hydrofoil vessels tend to be applied. Thanks to application of modern steam and gas turbines, water-jet propulsors are used successfully at large vessels, and design propulsive efficiency of such unit can reach 83% which is 11% more than that of the propeller screw.

Submarine application trends for water-jet propulsors are especially interesting, since it could increase the speed and decrease noise emissions of such crafts. However, such grand problems as viscous losses due to stream-outside flow vortex generation, and enormous flow friction in resultant eddy currents exist. Besides, increase of the speed of the submarine will inevitably entail cavitation rise at the pump blades which leads to fail. Application of supercavitation impeller blades could solve the problem.

The aerial propeller as a ship propulsor is installed at airfoil boats, amphibians, and aerostatic crafts. The operating principle of the last devices is based on pressurization under the ship bottom. The air bubble thereby created lifts the vessel over any surface (water, sand or boggy ground). With the same principle, air bubble methods can be different – central circular nozzle, nozzle ring with or without extension, labyrinth seal around the periphery of the bottom, etc.

THE NEW

The number of patents for new ship propulsors is very large. Thus, the water-jet propulsor was offered based on water heating. It consists of the contoured duct (water pipe) combined with the water heater. The water pipe is to be placed between hulls of the submarine. The sea inlet being located on the underside, the heated water is raised up via the free-flow duct and is ejected upper of the vessel contrarily to its direction. The water moves in the duct due to density difference. The main purpose of such propulsive unit is to provide noiseless movement of the submarine.

Well known propulsors working on the fishtail principle are also being retrofitted. Inventors state rather high (even comparing to propeller screws) efficiency and economy of such units. But they go behind on low speed of the experimental models. One of the Russian inventors offers equipment with additional special shaped bow propeller. The model equipped with such propeller and with a standard stern propeller demonstrated decent velocity gain – about 50% of the reference.

ELECTRICAL GENERATION

The electrical requirements of the training ship are supplied by the electrical plant. The electrical plant consists of three turbo-generators, one emergency generator, the emergency battery system, and the electrical distribution system. There are three 750 kW turbo-generators, arranged for individual and parallel operation with any two capable of carrying the maximum sea and port load of the ship and the third generator available for standby. Each turbo-generator is capable of a continuous rating of 750 kW, with a 25% overload for two hours.

A 200 kW emergency diesel generator is adequate for supplying power for emergency lighting, interior communication and power load requirements and also cold ship starting of the plant. Paralleling of main and emergency generators is prevented by an interlocking circuit which trips the bus tie breaker on the main switchboard and emergency switchboard when the generator breaker is closed.

One emergency battery bank, rated at 120 volts, 204 ampere-hours is provided for the ships emergency lighting, 120 V DC bus, fire door release system, diesel starting control circuit and the manual alarm and fire detection system. A recapitulation of data from Bender drawing 546 TV-001-301-06, electric load analysis, showing calculated loads for the turbo-generators, the emergency generator and the emergency batteries is shown below. It is noted that calculated values exceed the requirements encountered during sea trials.

LOAD SUMMARY TABULATION

Main Distribution System
Port Load 793.6 kW
Cruise Load 1372.9 kW
Training Load 1434.7 kW
Emergency Distribution System
Emergency Load 199.8 kW
Turbogenerator Rating
Continuous 750 kW
2 Hour Overload 938 kW
Diesel Generator Rating
Continuous 200 kW
Emergency Battery
Voltage120 Volts DC
Capacity 204 Ampere Hours

ELECTRICAL DISTRIBUTION SYSTEM

The electrical distribution system's main purpose is to distribute and control the supply of electrical power to all auxiliary and electrical equipment aboard ship. The electrical distribution system consists the cables, busses, circuit breakers, fueses, etc., necessary to safely distribute power throughout the ship.

MAIN SWITCHBOARD

The main switchboard, as the primary distribution center, distributes 450 volts, 3 phase, 60 cycle power. The 450 volt power is distributed to power panels throughout the ship and also to the emergency switchboard. Transformers at the lighting load centers step down the 450 volts to 120 volts. The lighting load centers distribute 120 volt power. There are three lighting load centers.

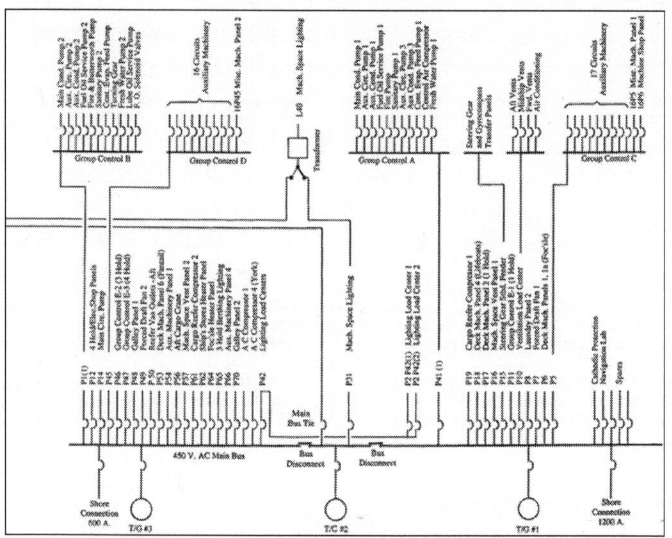

Fig. Main Electrical Power Distribution System.

The main switchboard is composed of three parts consisting of distribution panels at the right, three generator panels in the center, and distribution panels at the left (facing the front of the switchboard). Right side distribution and generator no.1 are connected to bus section 1, emergency switchboard feeder, machinery space lighting and generator no. 2 are connected to bus section 2. Left side distribution and generator no. 3 are connected to bus section 3. The three bus sections are normally connected together by removable disconnect bars.

EMERGENCY SWITCHBOARD

The emergency switchboard is the emergency power distribution center. The emergency switchboard distributes 450 volts and 120 volt power to auxiliaries which are vital under emergency conditions, essential lighting, safety and communication systems, and power necessary for dead ship starting.

Under normal operation, power for the emergency switchboard is supplied by the main generators through a bus-tie. In case of power failure, loss of voltage at the emergency switchboard will cause the emergency generator to automatically start. Simultaneously, the bus-tie will open disconnecting the main and the emergency switchboards. Then, the emergency generator breaker will close, energizing the emergency board and providing 200 kW of emergency power.

The emergency switchboard is composed of the following sections:

1. A generator and bus transfer section provides control for operation of the emergency generator and the bus transfer unit.
2. A distribution unit with a 450 volt, 60 cycle, 3 phase bus provides power for emergency power loads, and a 450/120 volt transformer bank, the secondary of which supplies a 120 volt, 3 phase, 60 cycle bus that provides power for emergency lighting and I.C. (Interior Communication) equipment.
3. A 120 volt D.C. panel for battery distribution.
4. An automatic charging unit is provided for charging the ship's emergency power battery.

The emergency switchboard is provided with the following instrumentation and controls.

1. Control switch with white indicator light for generator space heaters connected through the generator circuit breaker auxiliary switch and fed from circuit breakers on the 120 volts a.c. bus
2. A.C. ammeter with 5-ampere coil and 0-500 ampere scale
3. Three phase selector switch for above
4. A.C. voltmeter with 0-600 volt scale
5. Three phase selector switch for above
6. Polyphase indicating wattmeter with 0-300 kw scale
7. White indicating light to show power available from the emergency generator
8. Exciter field rheostat
9. Emergency generator voltage regulating equipment
10. White indicating light to show power available on the 450 volt bus from the main switchboard
11. Green indicator light to show diesel setup for automatic operation
12. Frequency meter with 55-65 cycle scale
13. Ground detector lights for the 450-volt, 3 phase generator
14. Ground circuit pushbutton switch with normally open contacts
15. Ground detector lights for the 120 volt, 3 phase final bus
16. Voltage regulator cutout switch

17. Frequency meter transfer switch, generator and bus

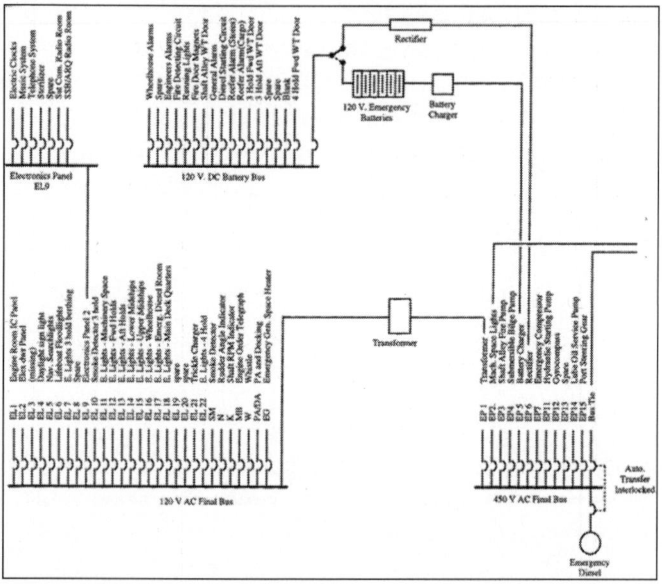

Fig. Emergency Electrical Power Distribution System.

SHORE POWER

When the ship is alongside, power can be supplied to the main switchboard from shore, through the shore connection. The original shore power connection was weather tight 600 ampere shore connection box is located at the after end of the house near the centerline, for supplying 440 volts, 3 phase, alternating current shore power to the main switchboard through a circuit breaker connection to the bus. Later, a 1200 amp connection box on the starboard side of the "B" deck passageway outside the engine room was installed to provide additional shore power capacity for when the ship is tied up in Buzzards Bay..

Provisions are made for synchronizing each main generator with the main bus. Thus, the main generators may be synchronized with the shore power for short periods while changing the load from one power source to another.

UNGROUNDED DISTRIBUTION SYSTEM

The electrical distribution system on board the Patriot State is referred to as an ungrounded distribution system. An ungrounded distribution system has no deliberate electrical connection to ground, ground being the hull of the ship. Its most significant advantage is that accidental contact between one hot line and ground (i.e. a ground fault) does not cause an outage (via circuit breaker tripping due to excessive current). Although a single ground fault on an ungrounded system does not cause an interruption in service, it is important that the fault be located and cleared immediately. If not remedied and a second ground fault occurs on any one of the other two phases, a short circuit current will result. This short circuit current may trip one or more circuit breakers. A double ground fault condition is shown in the next figure.

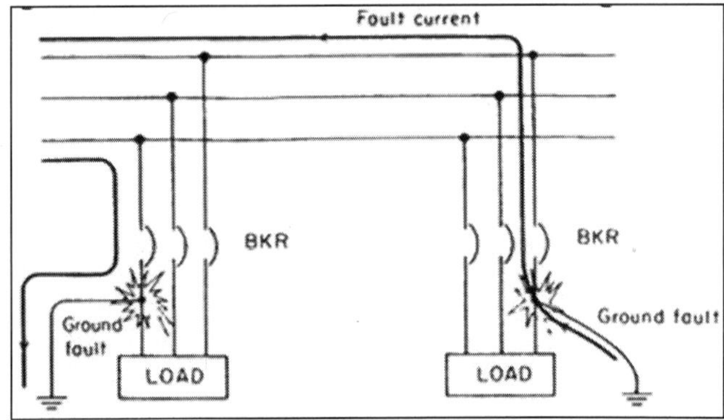

Fig. Path of current in an ungrounded distribution system resulting from grounds on two different phases

Ground fault detecting circuits for use in ungrounded systems to indicate the presence of a ground fault are shown below. Sub-figure (b) shows the ground fault detecting circuit on board the Patriot State. Three identical lamps are connected across the three phase line voltage. Each lamp has a resistor in series with it to limit the line current in the event of a single ground fault condition. The junction point of the three lamps is connected to ground (the hull) through a normally closed spring return switch. A low resistance ground fault on any one of the three hot lines will cause its respective lamp to burn dimly or even go out, (depending on the severity of the ground fault), with the other two lamps burning more brightly; with no ground faults all three lamps will be dim. The three lamps should have identical wattage ratings and a voltage rating equal to the line voltage. The normally closed spring return switch provides a means for comparing the normal and ground fault indications. Opening the spring return switch disconnects the ground fault detection circuit from the hull of the ship and all three lamps should return to a dim status.

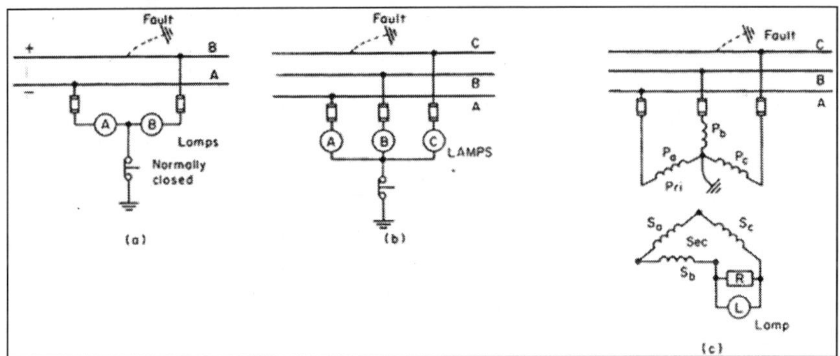

Fig. Ground Detection for ungrounded distribution systems: (a) Single Phase or DC (b) Three-phase, low voltage (c) Three-phase, high voltage.

Accidental ground faults should be repaired as soon as possible, for even one ground fault has a bad effect on the insulation. A single ground fault doubles the electrical stresses on the remaining insulation, thus increasing the possibility of an insulation breakdown in the other two phases.

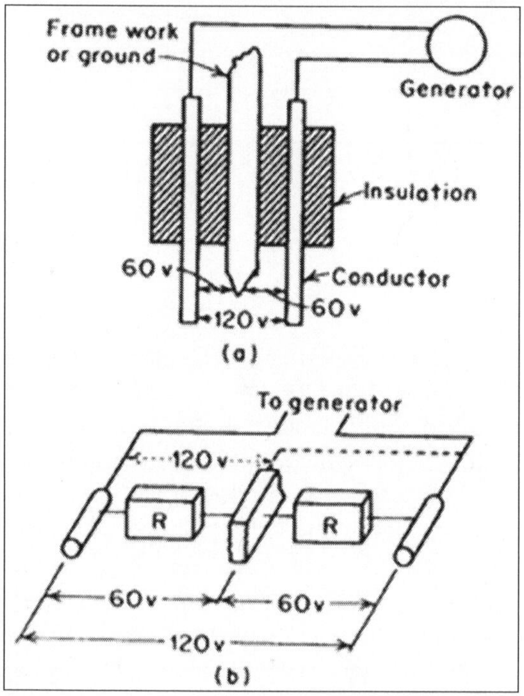

Fig. Distribution of Voltage between Conductors and Ground.

The doubled electrical stresses cause twice the electron leakage through the insulation, hastening the deterioration and shortening its life. The insulation resistance between each conductor and ground is R ohms. With 120 volts applied between the two conductors, the voltage between each conductor and ground is 60 volts. However, if one conductor has a ground fault condition, as indicated by the dotted line, the voltage difference between the other conductor and ground will rise to 120 volts. Hence the voltage stress on the insulation of the ungrounded conductor is doubled, and if a weak spot in the ungrounded conductor's insulation causes it to rupture, a short circuit will result.

TROUBLE SHOOTING UNGROUNDED DISTRIBUTION SYSTEMS

Troubles in ungrounded distribution systems usually manifest themselves by failure of a device to operate, an indication on the ground fault detection apparatus, smoke, or the overheating of a cable. Shorts and opens are relatively easy to locate and are indicated by blown fuses, tripped breakers, and voltage failure respectively. Grounds faults on the other hand, unless accompanied by a short or an open, are generally located by the process of elimination. The ground faulted circuit may be determined by opening breakers on the distribution panel one at a time, until the ground-detecting device on the switchboard indicates normal. Closing each breaker before opening the next keeps the interruption of service to a minimum. The opening of breakers supplying vital auxiliaries should be avoided until standby equipment is placed in operation. If this procedure fails either the ground fault is in the generator or more than one ground is present.

Multiple grounds faults may be located by opening the breakers one at a time, and leaving them open until the ground fault detecting lights indicate normal. Then with the ground faulted breaker left open, the other breakers should be closed until another ground fault is indicated. A ground fault in a generator may be identified by transferring the load to another machine and tripping the suspect machine from the line. If the generator has a ground fault,tripping it should return the ground detecting lights to normal. Tracking down the actual location of the ground faulted conductor is done best with a megohmmeter.

When doing so the breaker to the grounded faulted circuit should be blocked open, and a "Do Not Close-Man Working On Line" sign hung over the breaker. The megohmmeter method for tracking down a ground that may be in the supply cable, starter, motor cable, or in the motor itself. Always insure the circuit is dead before checking the insulation resistance. The supply cable may then be tested by applying a megohmmeter between the metal framework of the motor and the supply cable. A zero reading on the megohmmeter indicates a ground fault condition.

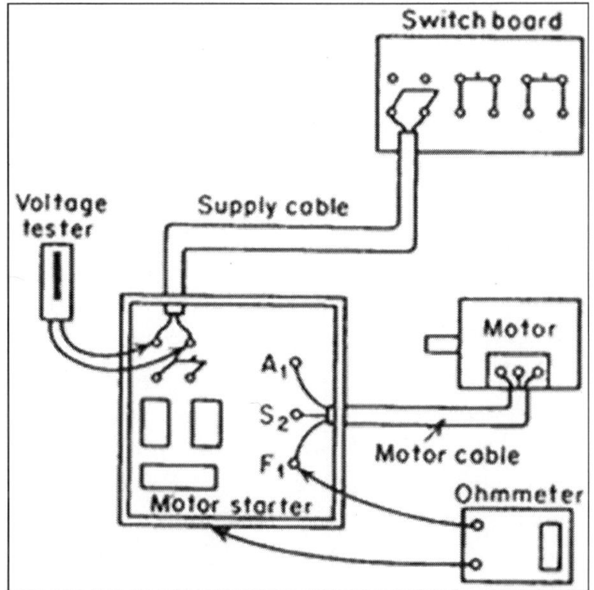

Fig. Tracking down a ground by the process of elimination.

GROUNDING ON THE 120 V. FINAL BUS

The 120 volt final bus provides the electrical distribution for lighting, appliances and also to the 120 volt electrical outlets (plugs). The pronged end of the outlet is the grounding connection . This insures that any appliance, power tool and anything else that plugs into the pronged end of the outlet is securely grounded to the ship's hull. This provides ground fault protection from high or low resistance ground faults in 120 volt a.c. appliances by providing a safe path for electricity in the event of a ground fault. This grounding circuit also allows the ground fault detecting equipment to detect the ground fault at the switchboard.

GROUNDING SUMMARY

All electrical equipment and outlets on board ship are securely grounded to the hull, so that in the event of ground faults, the ground lamps on the switchboard will indicate the fault. although all the electrical equipment on board is securely grounded to the hull, this type of distribution system is referred to as an ungrounded electrical distribution system. This is due to the fact that resistors in series with the ground detecting lamps prevent short circuit current, subsequent breakers tripping and power outages. The reason for this is obvious, you want a warning before loss of vital equipment. The ungrounded distribution system gives you this warning.

MARINE CURRENT POWER

Marine current power is a form of marine energy obtained from harnessing of the kinetic energy of marine currents, such as the Gulf stream. Although not widely used at present, marine current power has an important potential for future electricity generation. Marine currents are more predictable than wind and solar power. A 2006 report from United States Department of the Interior estimates that capturing just $^{1}/_{1,000th}$ of the available energy from the Gulf Stream would supply Florida with 35% of its electrical needs.

Strong ocean currents are generated from a combination of temperature, wind, salinity, bathymetry, and the rotation of the earth. The sun acts as the primary driving force, causing winds and temperature differences. Because there are only small fluctuations in current speed and stream location with minimal changes in direction, ocean currents may be suitable locations for deploying energy extraction devices such as turbines. Other effects such as regional differences in temperature and salinity and the Coriolis effectdue to the rotation of the earth are also major inuences. The kinetic energy of marine currents can be converted in much the same way that a wind turbine extracts energy from the wind, using various types of open-ow rotors.

The potential of electric power generation from marine tidal currents is enormous.

There are several factors that make electricity generation from marine currents very appealing when compared to other renewables:

- The high load factors resulting from the uid properties. The predictability of the resource, so that, unlike most of other renewables, the future availability of energy can be known and planned for.
- The potentially large resource that can be exploited with little environmental impact, thereby offering one of the least damaging methods for large-scale electricity generation.
- The feasibility of marine-current power installations to provide also base grid power, especially if two or more separate arrays with offset peak-ow periods are interconnected.

EARLY EXPERIENCES

The possible use of marine currents as an energy resource began to draw attention in the mid-1970s after the rst oil crisis. In 1974 several conceptual designs were presented at the MacArthur Workshop on Energy, and in 1976 the British General Electric Co. undertook a partially government-funded study which concluded that marine current power deserved more detailed research. Soon after, the ITD-Group in UK implemented a research program involving a year performance testing of a 3-m hydroDarrieus rotor deployed at Juba on the White Nile. The 1980s saw a number of small research projects to evaluate marine current power systems. The main countries where studies were carried out were the UK, Canada, and Japan. In 1992–1993 the Tidal Stream Energy Review identied specic sites in UK waters with suitable current speed to generate up to 58 TWh/year. It conrmed a total marine current power resource capable theoretically of meeting some 19% of the UK electricity demand.

In 1994–1995 the EU-JOULE CENEX project identified over 100 European sites ranging from 2 to 200 $km^2$of sea-bed area, many with power densities above 10 MW/km^2. Both the UK Government and the EU have committed themselves to internationally negotiated agreements designed to combat global warming. In order to comply with such agreements, an increase in large-scale electricity generation from renewable resources will be required. Marine currents have the potential to supply a substantial share of future EU electricity needs. The study of 106 possible sites for tidal turbines in the EU showed a total potential for power generation of about 50 TWh/year. If this resource is to be successfully utilized, the technology required could form the basis of a major new industry to produce clean power for the 21st century.

ALTERNATIVE TECHNOLOGIES IN MARINE-CURRENT-POWER APPLICATIONS

There are several types of open-ow devices that can be used in marine-current-power applications; many of them are modern descendants of the waterwheel or similar. However, the more technically sophisticated designs, derived from wind-power rotors, are the most likely to achieve enough cost-effectiveness and reliability to be practical in a massive marine-current-power future scenario. Even though there is no generally accepted term for these open-ow hydro-turbines, some sources refer to them as water-current turbines.

There are two main types of water current turbines that might be considered: axial-ow horizontal-axis propellers (with both variable-pitch or xed-pitch), and cross-ow vertical-axis Darrieus rotors. Both rotor types may be combined with any of the three main methods for supporting water-current turbines: oating moored systems, sea-bed mounted systems, and intermediate systems.

Sea-bed-mounted monopile structures constitute the rst-generation marine current power systems. They have the advantage of using existing (and reliable) engineering know-how, but they are limited to relatively shallow waters (about 20 to 40 m depth).

ENVIRONMENTAL EFFECTS

Ocean currents are instrumental in determining the climate in many regions around the world. While little is known about the effects of removing ocean current energy, the impacts of removing current energy on the farfield environment may be a significant environmental concern. The typical turbine issues with blade strike, entanglement of marine organisms, and acoustic effects still exists; however, these may be magnified due to the presence of more diverse populations of marine organisms using ocean currents formigration purposes. Locations can be further offshore and therefore require longer power cables that could affect the marine environment with electromagnetic output. The Tethys database provides access to scientific literature and general information on the potential environmental effects of ocean current energy.

TURBO GENERATORS

The primary electrical power requirements of the Patriot State are met by three, alternating current, geared steam turbine, ships service generators. All three ships service generators are located in the main machinery space, between frames 118 and 124 on the starboard side of the auxiliary machinery flat. All three generators are identical, all having the same characteristics and ratings. Each of the ships service generators are rated to produce 750 kw continuous, 450 volts, 0.8 power factor, 60 cycle, 3 phase electrical power.

The three ships service generators also have the same overload ratings. Each generator is capable of operating at a 25% overload (938 kw) for a period of two hours without overheating. Also, the design back pressure for auxiliary turbines, that drive the ships service generator, is 28" hg vacuum. But there is no loss of normal generator capacity with a back pressure as high as 20" hg vacuum. The three ships service generators are arranged for individual, or parallel operation. Provisions are also made for the ships service generators to be paralleled with shore power. This will facilitate the transfer from shore power to ships service generators. During parallel operation of the ships service generators, any two are capable of supplying the maximum sea or port load. The third ship service generator is therefore free from maintenance or available as a standby unit.

TURBO GENERATOR SPECIFICATIONS

Type:
DRV-618M, 6-stage 11108/2300 RPM
Rating:
Continuous 750kW, 0.8PF, 450 volts
60 cycle 3 phase AC
1200 RPM
Overload 25% for two hours
Rated Steam Condition:
Inlet steam pressure 600 psig
Initial steam pressure 900° F.
Exhaust pressure 2" Hg abs.

CONSTRUCTION

The turbine and quill shaft of the gear pinion are rigidly connected. The pinion and turbine shaft are flexibly coupled with a dental type coupling. The turbine shaft and the gear pinion are supported by four bearings, three in the reduction gear casing and one at the exhaust end of the turbine. The gear end of the turbine shaft is supported by one of the bearings in the gear casing.

The exhaust end of the turbine casing is carried by flexible standards mounted on the structural steel base. These supports provide axial flexibility, thereby, permitting the turbine casing and shaft to expand with temperature change in the direction parallel to the shaft. The supports are rigid in the direction perpendicular to the shaft. The turbine casing is supported at the high-pressure or gear end by the reduction-gear casing. Steam enters through a steam strainer and in turn passes through the throttle valve and the controlling valves, then to the first-stage nozzleplate. The steam expands in successive stages from initial pressure to final or condenser pressure.

STEAM SEAL SYSTEM

The two shaft packing boxes are connected into a system that provides means for disposing of any excess amount of steam that leaks past the packing rings and also for supplying sealing steam to prevent leakage of air into the turbine. The two valves in the steam-seal system, if operated properly, permit the maintenance of a sealing pressure of one to two psig at the shaft packing boxes. the steam for gland sealing comes form the auxiliary exhaust system.

Since the turbine is a condensing unit, a vacuum will exist in the entire turbine shell during periods of starting and stopping, and of any other time when the turbine is operating at low load. It is during such times, that sealing steam is required to prevent air leaking into the turbine. During periods of normal full load, there is enough leakage from the high pressure end to seal the low pressure. Therefore, no sealing-steam is needed.

TURBINE SHAFT BEARING

The bearing supports the low pressure end of the turbine shaft. The gear end of the turbine shaft is supported by the pinion shaft coupling. The turbine shaft bearing is ball seated for self alignment and is babbitt lined. It is split horizontally to facilitate assembly and is clamped to its seat by a bearing cap. Oil enters the bearing surface at the horizontal joint where the babbitt is relieved to permit the journal to draw the oil around the shaft in the direction of rotation. The oil passes at through drain holes that are drilled into grooves at the ends of the bearing.

GOVERNOR AND OVERSPEED TRIP

The governing system is designed to automatically maintain narrow speed regulation within the normal operating range under varying load and steam conditions; and provides a means for adjusting the speed and for changing the portion of total load each set is to carry when running in parallel with another set. It also includes an emergency device that closes the throttle valve as the turbine overspeeds.

The governor is driven through a spiral gear, the spiral pinion being directly connected to the shaft of the low-speed member of the reduction gear and thereby driving the governor at a speed that is directly proportional to the speed of the turbine. With an increase in turbine speed, the governor weights move outward and draw the pilot valve downward with a reduction in turbine speed, the weights move in and push the pilot valve upward. It is the position of the pilot valve with respect to the pilot valve bushing that determines the flow of oil to the operating cylinder and therefore the position of the operating valves (governor valves) that admit steam to the turbine. The emergency tripping mechanism forms the connecting link between the emergency governor and the emergency feature of the throttle valve.

When overspeeding occurs, the emergency governor operates and strikes the trigger which turns on its pivot pin and unlatches the trip rod. This rod is pushed downward immediately by the compression spring, swinging the bell crank around its pivot pin. The bell crank is connected to the valve and stem of the oil release valve. When the bell crank swings about, its axis resulting from the mechanism being tripped by hand by striking the trip button or by the emergency governor, the oil release valve is lifted from its seat in the body allowing oil to drain from under the throttle valve operating cylinder through the pipe line to the valve body and from there into the bearing bracket. This in turn will result in a quick closing of the throttle valve thereby shutting off steam flow to the turbine.

To reset the tripping device after it has operated, first reset the throttle valve. This will raise the cylinder position in the throttle valve operating cylinder above the oil by-pass slots. Reset the tripping device by rotating the bell crank toward the oil release valve until the trip rod is relatched. Now with oil pressure being applied under the piston by means of the hand oil pump, the throttle valve can again be opened allowing steam to pass into the turbine. Care should be taken not to reset the emergency device until the turbine speed has dropped below 7800 RPM, otherwise the emergency governor plunger may hit the trigger. In addition to the emergency governor, there is a back pressure trip installed on the number 1 and 2 gererators which protects the turbine against excessive exhaust pressure. This device trips at 5 psig exhaust pressure. Each turbine is also fitted with a sentinel valve which relieves at 2 psig exhaust pressure.

LUBE OIL SYSTEM

The shaft driven oil pump draws oil from the tank located in the base of the gear casing and delivers it through the two separate systems to the hydraulic governing mechanism and to the bearing and gear mesh lubricating system. Oil to the lubricating system passes through the oil strainer, manifold and oil cooler. The hand oil pump likewise draws oil from the tank but delivers it directly to the bearings and hydraulic mechanism.

The normal oil pressure delivered to the governing mechanism should be approximately 55 psig. The normal bearing oil pressure should be approximately 8 psig. A fair running temperature of the return oil is 140-160° F. The maximum temperature rise at ay bearing should not exceed 50° F. The lube oil strainer is of the duplex type. It consists of two cylindrical strainer baskets connected by a manifold with a plug type valve for delivering the olil through one or the other of the two strainers. While one strainer is in use, the other may be removed for cleaning.

The lube oil system is equipped with a low lube oil alarm. The alarm sounds below 4 psig and is reset at 6 psig. There is also a low oil presssure switch that activates to open th main generator circuit breaker when the throttle valve hydraulic line pressure drops below 15 psig. This switch resets at 30 psig.

REDUCTION GEARS

The type 3-387 reduction gear is a single-reduction, single-helical type and reduces the turbine speed of 9963 rpm to the generator speed of 1200 rpm.

STEAM TURBINE

A steam turbine is a device which extracts thermal energy from pressurized steam and uses it to do mechanical work on a rotating output shaft. Its modern manifestation was invented by Sir Charles Parsons in 1884. Because the turbine generates rotary motion, it is particularly suited to be used to drive an electrical generator – about 90% of all electricity generation in the United States (1996) is by use of steam turbines. The steam turbine is a form of heat engine that derives much of its improvement in thermodynamic efficiency from the use of multiple stages in the expansion of the steam, which results in a closer approach to the ideal reversible expansion process.

Fig. The rotor of a modern steam turbine used in a power plant.

HISTORY

A 250 kW industrial steam turbine from 1910 (right) directly linked to agenerator (left).

The first device that may be classified as a reaction steam turbine was little more than a toy, the classic Aeolipile, described in the 1st century by

Greek mathematician Hero of Alexandria in Roman Egypt. In 1551, Taqi al-Din in Ottoman Egypt described a steam turbine with the practical application of rotating a spit. Steam turbines were also described by the Italian Giovanni Branca (1629) andJohn Wilkins in England (1648). The devices described by Taqi al-Din and Wilkins are today known as steam jacks.

The modern steam turbine was invented in 1884 by Sir Charles Parsons, whose first model was connected to a dynamo that generated 7.5 kW (10 hp) of electricity. The invention of Parsons' steam turbine made cheap and plentiful electricity possible and revolutionized marine transport and naval warfare. Parsons' design was a reaction type.

His patent was licensed and the turbine scaled-up shortly after by an American, George Westinghouse. The Parsons turbine also turned out to be easy to scale up. Parsons had the satisfaction of seeing his invention adopted for all major world power stations, and the size of generators had increased from his first 7.5 kW set up to units of 50,000 kW capacity. Within Parson's lifetime, the generating capacity of a unit was scaled up by about 10,000 times, and the total output from turbo-generators constructed by his firm C. A. Parsons and Company and by their licensees, for land purposes alone, had exceeded thirty million horse-power.

A number of other variations of turbines have been developed that work effectively with steam. The de Laval turbine (invented by Gustaf de Laval) accelerated the steam to full speed before running it against a turbine blade. De Laval's impulse turbine is simpler, less expensive and does not need to be pressure-proof.

It can operate with any pressure of steam, but is considerably less efficient. fr:Auguste Rateau developed a pressure compounded impulse turbine using the de Laval principle as early as 1896,obtained a US patent in 1903, and applied the turbine to a French torpedo boat in 1904. He taught at the École des mines de Saint-Étienne for a decade until 1897, and later founded a successful company that was incorporated into the Alstom firm after his death. One of the founders of the modern theory of steam and gas turbines was Aurel Stodola, a Slovak physicist and engineer and professor at the Swiss Polytechnical Institute (now ETH) in Zurich. His work Die Dampfturbinen und ihre Aussichten als Wärmekraftmaschinen (English: The Steam Turbine and its prospective use as a Mechanical Engine) was published in Berlin in 1903. A further book Dampf und Gas-Turbinen(English: Steam and Gas Turbines) was published in 1922.

The Brown-Curtis turbine, an impulse type, which had been originally developed and patented by the U.S. company International Curtis Marine Turbine Company, was developed in the 1900s in conjunction with John Brown & Company. It was used in John Brown-engined merchant ships and warships, including liners and Royal Navy warships.

MANUFACTURING

The manufacturing industry for steam turbines is dominated by Chinese power equipment makers. Harbin Electric, Shanghai Electric, and Dongfang Electric, the top three power equipment makers in China collectively hold a majority stake in the worldwide market share for steam turbines in 2009-10 according to Platts. Other manufacturers with minor market share include Bhel, Siemens, Alstom, GE, Mitsubishi Heavy Industries, and Toshiba. The consulting firm Frost & Sullivan projects that manufacturing of steam turbines will become more consolidated by 2020 as Chinese power manufacturers win increasing business outside of China.

TYPES

Steam turbines are made in a variety of sizes ranging from small <0.75 kW (<1 hp) units (rare) used as mechanical drives for pumps, compressors and other shaft driven equipment, to 1 500 000 kW (1.5 GW; 2 000 000 hp) turbines used to generate electricity. There are several classifications for modern steam turbines.

Blade and Stage Design

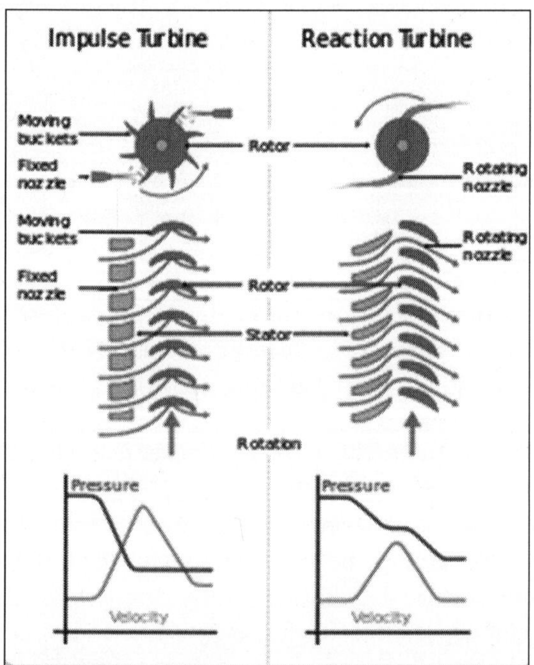

Fig. Schematic diagram outlining the difference between an impulse and a 50% reaction turbine.

Turbine blades are of two basic types, blades and nozzles. Blades move entirely due to the impact of steam on them and their profiles do not converge.

This results in a steam velocity drop and essentially no pressure drop as steam moves through the blades. A turbine composed of blades alternating with fixed nozzles is called an impulse turbine, Curtis turbine, Rateau turbine, or Brown-Curtis turbine. Nozzles appear similar to blades, but their profiles converge near the exit. This results in a steam pressure drop and velocity increase as steam moves through the nozzles. Nozzles move due to both the impact of steam on them and the reaction due to the high-velocity steam at the exit. A turbine composed of moving nozzles alternating with fixed nozzles is called a reaction turbineor Parsons turbine. Except for low-power applications, turbine blades are arranged in multiple stages in series, called compounding, which greatly improves efficiency at low speeds. A reaction stage is a row of fixed nozzles followed by a row of moving nozzles. Multiple reaction stages divide the pressure drop between the steam inlet and exhaust into numerous small drops, resulting in a pressure-compounded turbine. Impulse stages may be either pressure-compounded, velocity-compounded, or pressure-velocity compounded. A pressure-compounded impulse stage is a row of fixed nozzles followed by a row of moving blades, with multiple stages for compounding. This is also known as a Rateau turbine, after its inventor. A velocity-compounded impulse stage (invented by Curtis and also called a "Curtis wheel") is a row of fixed nozzles followed by two or more rows of moving blades alternating with rows of fixed blades. This divides the velocity drop across the stage into several smaller drops. A series of velocity-compounded impulse stages is called a pressure-velocity compounded turbine.

By 1905, when steam turbines were coming into use on fast ships (such asHMS Dreadnought) and in land-based power applications, it had been determined that it was desirable to use one or more Curtis wheels at the beginning of a multi-stage turbine (where the steam pressure is highest), followed by reaction stages. This was more efficient with high-pressure steam due to reduced leakage between the turbine rotor and the casing. This is illustrated in the drawing of the German 1905 AEG marine steam turbine.

The steam from the boilers enters from the right at high pressure through a throttle, controlled manually by an operator (in this case a sailor known as the throttleman). It passes through five Curtis wheels and numerous reaction stages (the small blades at the edges of the two large rotors in the middle) before exiting at low pressure, almost certainly to a condenser. The condenser provides a vacuum that maximizes the energy extracted from the steam, and condenses the steam into feedwater to be returned to the boilers. On the left are several additional reaction stages (on two large rotors) that rotate the turbine in reverse for astern operation, with steam admitted by a separate throttle. Since ships are rarely operated in reverse, efficiency is not a priority in astern turbines, so only a few stages are used to save cost.

Blade Design Challenges

A major challenge facing turbine design is reducing the creep experienced by the blades. Because of the high temperatures and high stresses of operation, steam turbine materials become damaged through these mechanisms. As temperatures are increased in an effort to improve turbine efficiency, creep becomes more significant. To limit creep, thermal coatings and superalloys with solid-solution strengthening and grain boundary strengthening are used in blade designs.

Protective coatings are used to reduce the thermal damage and to limit oxidation. These coatings are often stabilized zirconium dioxide-based ceramics. Using a thermal protective coating limits the temperature exposure of the nickel superalloy. This reduces the creep mechanisms experienced in the blade. Oxidation coatings limit efficiency losses caused by a buildup on the outside of the blades, which is especially important in the high-temperature environment.

The nickel-based blades are alloyed with aluminum and titanium to improve strength and creep resistance. The microstructure of these alloys is composed of different regions of composition. A uniform dispersion of the gamma-prime phase – a combination of nickel, aluminum, and titanium – promotes the strength and creep resistance of the blade due to the microstructure. Refractory elements such as rhenium and ruthenium can be added to the alloy to improve creep strength. The addition of these elements reduces the diffusion of the gamma prime phase, thus preserving the fatigue resistance, strength, and creep resistance.

Steam Supply and Exhaust Conditions

Fig. A low-pressure steam turbine in a nuclear power plant. These turbines exhaust steam at a pressure below atmospheric.

These types include condensing, non-condensing, reheat, extraction and induction. Condensing turbines are most commonly found in electrical power plants. These turbines receive steam from a boiler and exhaust it to acondenser. The exhausted steam is at a pressure well below atmospheric, and is in a partially condensed state, typically of a quality near 90%.

Non-condensing or back pressure turbines are most widely used for process steam applications. The exhaust pressure is controlled by a regulating valve to suit the needs of the process steam pressure. These are commonly found at refineries, district heating units, pulp and paper plants, and desalination facilities where large amounts of low pressure process steam are needed.

Reheat turbines are also used almost exclusively in electrical power plants. In a reheat turbine, steam flow exits from a high pressure section of the turbine and is returned to the boiler where additional superheat is added. The steam then goes back into an intermediate pressure section of the turbine and continues its expansion. Using reheat in a cycle increases the work output from the turbine and also the expansion reaches conclusion before the steam condenses, thereby minimizing the erosion of the blades in last rows. In most of the cases, maximum number of reheats employed in a cycle is 2 as the cost of super-heating the steam negates the increase in the work output from turbine.

Extracting type turbines are common in all applications. In an extracting type turbine, steam is released from various stages of the turbine, and used for industrial process needs or sent to boiler feedwater heaters to improve overall cycle efficiency. Extraction flows may be controlled with a valve, or left uncontrolled. Induction turbines introduce low pressure steam at an intermediate stage to produce additional power.

Casing or Shaft Arrangements

These arrangements include single casing, tandem compound and cross compound turbines. Single casing units are the most basic style where a single casing and shaft are coupled to a generator. Tandem compound are used where two or more casings are directly coupled together to drive a single generator. A cross compound turbine arrangement features two or more shafts not in line driving two or more generators that often operate at different speeds. A cross compound turbine is typically used for many large applications.

Two-Flow Rotors

The moving steam imparts both a tangential and axial thrust on the turbine shaft, but the axial thrust in a simple turbine is unopposed. To maintain the correct rotor position and balancing, this force must be counteracted by an opposing force. Thrust bearings can be used for the shaft bearings, the rotor can use dummy pistons, it can be double flow- the steam enters in the middle of the shaft and exits at both ends, or a combination of any of these. In a double flow rotor, the blades in each half face opposite ways, so that the axial forces negate each other but the tangential forces act together. This design of rotor is also called two-flow, double-axial-flow, or double-exhaust. This arrangement is common in low-pressure casings of a compound turbine.

Fig. A two-flow turbine rotor. The steam enters in the middle of the shaft, and exits at each end, balancing the axial force.

PRINCIPLE OF OPERATION AND DESIGN

Fig. A simple turbine schematic of the Parsons type: rotating and fixed stators alternate and steam pressure drops by a fraction of the total across each pair. The stators grow larger as pressure drops.

An ideal steam turbine is considered to be an isentropic process, or constant entropy process, in which the entropy of the steam entering the turbine is equal to the entropy of the steam leaving the turbine. No steam turbine is truly isentropic, however, with typical isentropic efficiencies ranging from 20–90% based on the application of the turbine. The interior of a turbine comprises several sets of blades orbuckets. One set of stationary blades is connected to the casing and one set of rotating blades is connected to the shaft. The sets intermesh with certain minimum clearances, with the size and configuration of sets varying to efficiently exploit the expansion of steam at each stage.

Turbine Efficiency

To maximize turbine efficiency the steam is expanded, doing work, in a number of stages. These stages are characterized by how the energy is extracted from them and are known as either impulse or reaction turbines. Most steam turbines use a mixture of the reaction and impulse designs: each stage behaves as either one or the other, but the overall turbine uses both. Typically, higher pressure sections are reaction type and lower pressure stages are impulse type.

Impulse Turbines

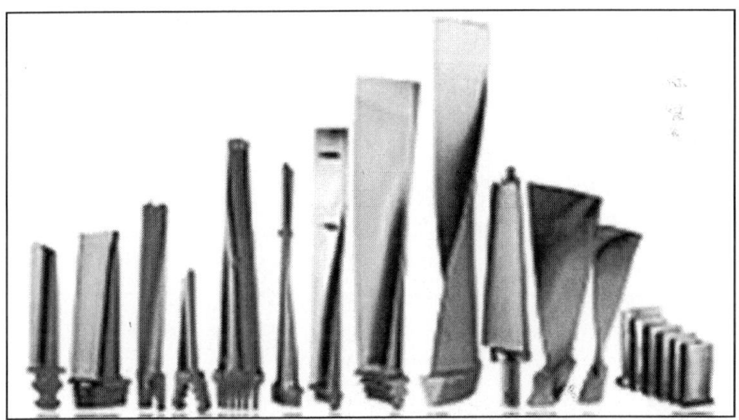

Fig. A selection of impulse turbine blades.

An impulse turbine has fixed nozzles that orient the steam flow into high speed jets. These jets contain significant kinetic energy, which is converted into shaft rotation by the bucket-like shaped rotor blades, as the steam jet changes direction. A pressure drop occurs across only the stationary blades, with a net increase in steam velocity across the stage. As the steam flows through the nozzle its pressure falls from inlet pressure to the exit pressure (atmospheric pressure, or more usually, the condenser vacuum). Due to this high ratio of expansion of steam, the steam leaves the nozzle with a very high velocity. The steam leaving the moving blades has a large portion of the maximum velocity of the steam when leaving the nozzle. The loss of energy due to this higher exit velocity is commonly called the carry over velocity or leaving loss.

The law of moment of momentum states that the sum of the moments of external forces acting on a fluid which is temporarily occupying the control volume is equal to the net time change of angular momentum flux through the control volume. The swirling fluid enters the control volume at radius r_1 with tangential velocity Vw_1 and leaves at radius r_2 with tangential velocity Vw_2.

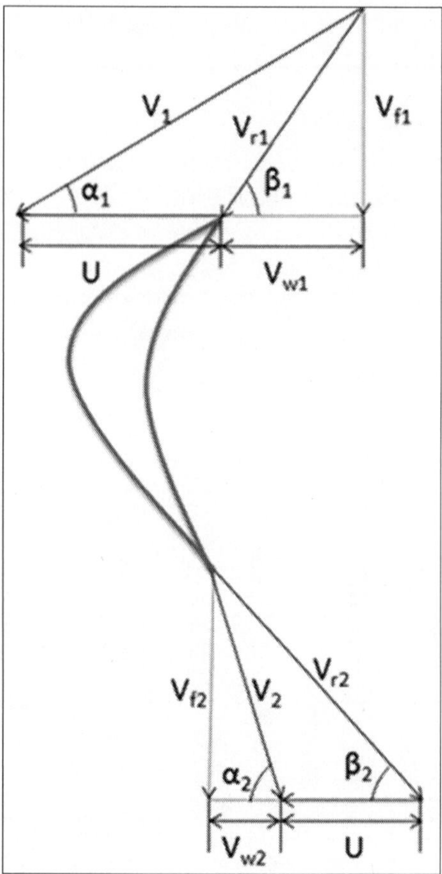

Fig. Velocity triangle.

A velocity triangle paves the way for a better understanding of the relationship between the various velocities. In the adjacent figure we have:

V_1 and V_2 are the absolute velocities at the inlet and outlet respectively.

V_{f1} and V_{f2} are the flow velocities at the inlet and outlet respectively.

V_{w1}+U and V_{w2} are the swirl velocities at the inlet and outlet respectively.

V_{r1} and V_{r2} are the relative velocities at the inlet and outlet respectively.

U_1 and U_2 are the velocities of the blade at the inlet and outlet respectively.

α is the guide vane angle and β is the blade angle.

Then by the law of moment of momentum, the torque on the fluid is given by:

$$T = \dot{m}(r_2 V_{w2} - r_1 V_{w1})$$

For an impulse steam turbine: $r_2 = r_1 = r$. Therefore, the tangential force on the blades is $F_u = \dot{m}(V_{w1} - V_{w2})$. The work done per unit time or power developed: $W = T * \omega$. When ω is the angular velocity of the turbine, then the blade speed is $U = \omega * r$. The power developed is then $W = \dot{m}U(\Delta V_w)$.

Blade Efficiency

Blade efficiency (η_b) can be defined as the ratio of the work done on the blades to kinetic energy supplied to the fluid, and is given by

$$\eta_b = \frac{Work\ Done}{Kinetic\ Energy\ Supplied} = \frac{2UV_w}{V_1^2}$$

Stage Efficiency

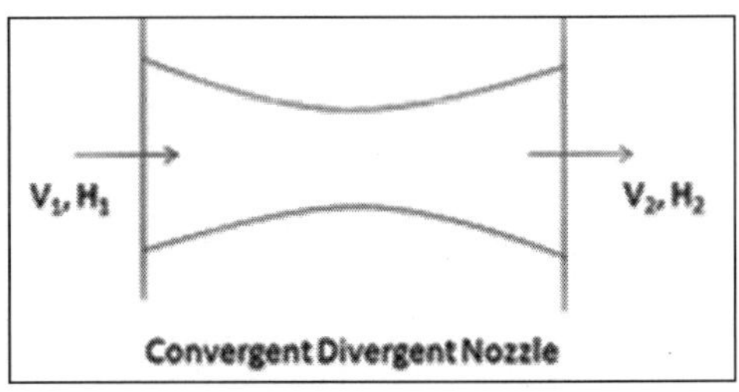

Fig. Convergent-divergent nozzle.

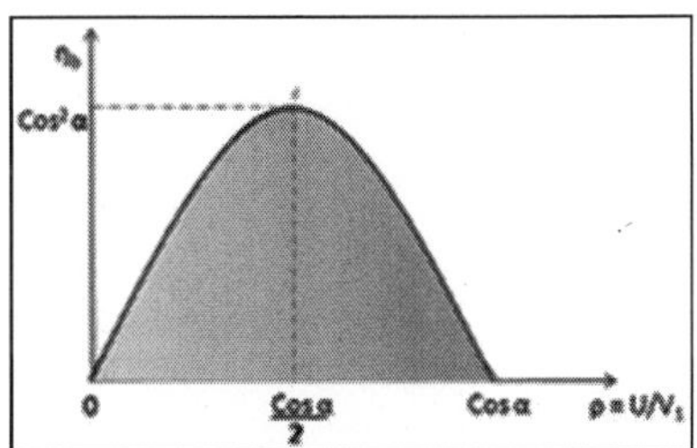

Fig. Graph depicting efficiency of Impulse turbine.

A stage of an impulse turbine consists of a nozzle set and a moving wheel. The stage efficiency defines a relationship between enthalpy drop in the nozzle and work done in the stage.

$$\eta_{stage} = \frac{Work\ done\ on\ blade}{Energy\ supplied\ per\ stage} = \frac{U\Delta V_w}{\Delta h}$$

Where $\Delta h = h_2 - h_1$ is the specific enthalpy drop of steam in the nozzle.

By the first law of thermodynamics: $h_1 + \frac{V_1^2}{2} = h_2 + \frac{V_2^2}{2}$

Assuming that V_1 is appreciably less than V_2, we get $\Delta h \approx \frac{V_2^2}{2}$ Furthermore, stage efficiency is the product of blade efficiency and nozzle efficiency, or $\eta_{stage} = \eta_b * \eta_N$

Nozzle efficiency is given by $\eta_N = \frac{V_2^2}{2(h_1 - h_2)}$, where the enthalpy (in J/Kg) of steam at the entrance of the nozzle is h_1 and the enthalpy of steam at the exit of the nozzle is h_2. $\Delta V_w = V_{w1} - (-V_{w2})$ $\Delta V_w = V_{w1} + V_{w2}$.

$$\Delta V_w = V_{r1}\cos\beta_1 + V_{r2}\cos\beta_2$$
$$\Delta V_w = V_{r1}\cos\beta_1\left(1 + \frac{V_{r2}\cos\beta_2}{V_{r1}\cos\beta_1}\right)*$$

The ratio of the cosines of the blade angles at the outlet and inlet can be taken and denoted $c = \frac{\cos\beta_2}{\cos\beta_1}$. The ratio of steam velocities relative to the rotor speed at the outlet to the inlet of the blade is defined by the friction coefficient $k = \frac{V_{r2}}{V_{r1}}$.

k<1 and depicts the loss in the relative velocity due to friction as the steam flows around the blades (k=1 for smooth blades).

$$\eta_b = \frac{2U\Delta V_w}{V_1^2} = \frac{2U(\cos\alpha_1 - U/V_1)(1+kc)}{V_1}$$

The ratio of the blade speed to the absolute steam velocity at the inlet is termed the blade speed ratio $\rho = \frac{U}{V_1}$

η_b is maximum when $\frac{d\eta_b}{d\rho} = 0$ or, $\frac{d}{d\rho}(2\cos\alpha_1 - \rho^2(1+kc)) = 0$. That implies $\rho = \frac{\cos\alpha_1}{2}$ and therefore $\frac{U}{V_1} = \frac{\cos\alpha_1}{2}$. Now $\rho_{opt} = \frac{U}{V_1} = \frac{\cos\alpha_1}{2}$ (for a single stage impulse turbine)

Therefore the maximum value of stage efficiency is obtained by putting the value of $\frac{U}{V_1} = \frac{\cos\alpha_1}{2}$ in the expression of η_b/

We get: $(\eta_b)_{max} = 2(\rho\cos\alpha_1 - \rho^2)(1+kc) = \frac{\cos^2\alpha_1(1+kc)}{2}$.

For equiangular blades, $\beta_1 = \beta_2$, therefore c=1, and we get $(\eta_b)_{max} = \frac{\cos^2\alpha_1(1+k)}{2}$. If the friction due to the blade surface is neglected then $(\eta_b)_{max} = \cos^2\alpha_1$.

Conclusions on maximum efficiency

$$(\eta_b)_{max} = \cos^2\alpha_1$$

1. For a given steam velocity work done per kg of steam would be maximum when $\cos^2\alpha_1 = 1$ or $\alpha_1 = 0$.
2. As α_1 increases, the work done on the blades reduces, but at the same time surface area of the blade reduces, therefore there are less frictional losses.

Reaction Turbines

In the reaction turbine, the rotor blades themselves are arranged to form convergent nozzles. This type of turbine makes use of the reaction force produced as the steam accelerates through the nozzles formed by the rotor. Steam is directed onto the rotor by the fixed vanes of the stator. It leaves the stator as a jet that fills the entire circumference of the rotor. The steam then changes direction and increases its speed relative to the speed of the blades. A pressure drop occurs across both the stator and the rotor, with steam accelerating through the stator and decelerating through the rotor, with no net change in steam velocity across the stage but with a decrease in both pressure and temperature, reflecting the work performed in the driving of the rotor.

Blade Efficiency

Energy input to the blades in a stage:

E = Δh is equal to the kinetic energy supplied to the fixed blades (f)+ the kinetic energy supplied to the moving blades (m).

Or, E = enthalpy drop ovcr thc fixed blades, Δhf + enthalpy drop over the moving blades, Δh_m.

The effect of expansion of steam over the moving blades is to increase the relative velocity at the exit. Therefore the relative velocity at the exit V_{r2} is always greater than the relative velocity at the inlet V_{r1}.

In terms of velocities, the enthalpy drop over the moving blades is given by:

$\Delta h_m = \dfrac{V_{r2}^2 - V_{r1}^2}{2}$ (it contributes to a change in static pressure) The enthalpy drop in the fixed blades, with the assumption that the velocity of steam entering the fixed blades is equal to the velocity of steam leaving the previously moving blades is given by:

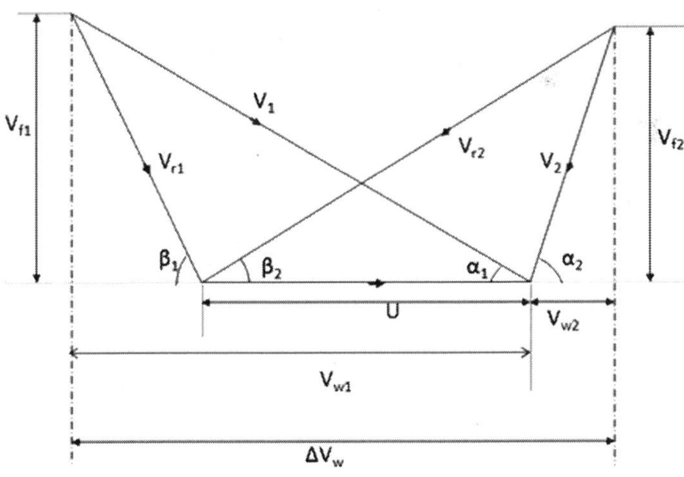

Fig. Velocity diagram.

$\Delta h_f = \dfrac{V_1^2 - V_0^2}{2}$ where V_0 is the inlet velocity of steam in the nozzle $V_0$is very small and hence can be neglected Therefore, $\Delta h_f = \dfrac{V_1^2}{2}$

$$E = \Delta h_f + \Delta h_m$$

$$E = \frac{V_1^2}{2} + \frac{V_{r2}^2 - V_{r1}^2}{2}$$

A very widely used design has half degree of reaction or 50% reaction and this is known as Parson's turbine. This consists of symmetrical rotor and stator blades. For this turbine the velocity triangle is similar and we have:

$$\alpha_1 = \beta_{2,}\ \beta_1 = \alpha_2$$

$$V_1 = V_{r2,}\ V_{r1} = V_2$$

Assuming Parson's turbine and obtaining all the expressions we get

$$E = V_1^2 - \frac{V_{r1}^2}{2}$$

From the inlet velocity triangle we have $V_{r1}^2 = V_1^2 + U^2 - 2UV_1\cos\alpha_1$

$$E = V_1^2 - \frac{V_1^2}{2} - \frac{U^2}{2} + \frac{2UV_1\cos\alpha_1}{2}$$

$$E = \frac{V_1^2 - U^2 + 2UV_1\cos\alpha_1}{2}$$

Work done (for unit mass flow per second): $W = U * \Delta V_w = U * (2 * V_1\cos\alpha_1 - U)$
Therefore the blade efficiency is given by

$$\eta_b = \frac{2U(2V_1\cos\alpha_1 - U)}{V_1^2 - U^2 + 2V_1U\cos\alpha_1}$$

Condition of Maximum Blade Efficiency

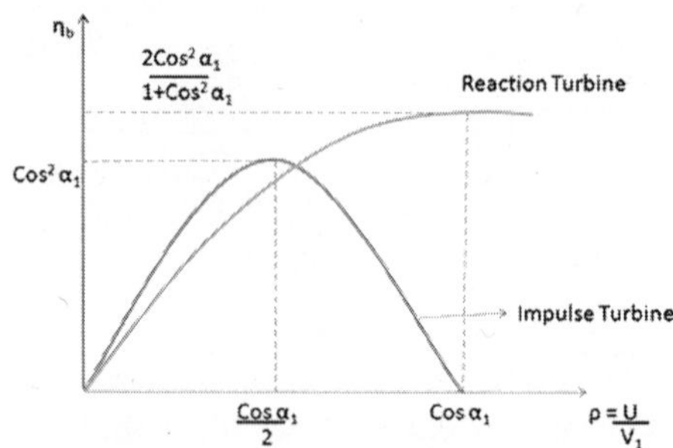

Fig. Comparing Efficiencies of Impulse and Reaction turbines.

If $\rho = \frac{U}{V_1}$, then $(\eta_b)_{max} = \frac{2\rho(\cos\alpha_1 - \rho)}{V_1^2 - U^2 + 2UV_1\cos\alpha_1}$ For maximum efficiency $\frac{d\eta_b}{d\rho} = 0$, we get

$$(1 - \rho^2 + 2\rho\cos\alpha_1)(4\cos\alpha_1 - 4\rho) - 2\rho(2\cos\alpha_1 - \rho)(-2\rho + 2\cos\alpha_1) = 0$$

and this finally gives $\rho_{opt} = \frac{U}{V_1} = \cos\alpha_1$ Therefore $(\eta_b)_{max}$ is found by putting the value of $\rho = \cos\alpha_1$ in the expression of blade efficiency

$$(\eta_b)_{reaction} = \frac{2\cos^2\alpha_1}{1 + \cos^2\alpha_1}$$

$$(\eta_b)_{impulse} = \cos^2\alpha_1$$

Operation and Maintenance

Fig. A modern steam turbine generator installation.

Because of the high pressures used in the steam circuits and the materials used, steam turbines and their casings have high thermal inertia. When warming up a steam turbine for use, the main steam stop valves (after the boiler) have a bypass line to allow superheated steam to slowly bypass the valve and proceed to heat up the lines in the system along with the steam turbine. Also, a turning gear is engaged when there is no steam to slowly rotate the turbine to ensure even heating to prevent uneven expansion. After first rotating the turbine by the turning gear, allowing time for the rotor to assume a straight plane (no bowing), then the turning gear is disengaged and steam is admitted to the turbine, first to the astern blades then to the ahead blades slowly rotating the turbine at 10–15 RPM (0.17–0.25 Hz) to slowly warm the turbine. The warm up procedure for large steam turbines may exceed ten hours.

During normal operation, rotor imbalance can lead to vibration, which, because of the high rotation velocities, could lead to a blade breaking away from the rotor and through the casing. To reduce this risk, considerable efforts are spent to balance the turbine. Also, turbines are run with high quality steam: either superheated (dry) steam, or saturated steam with a high dryness fraction. This prevents the rapid impingement and erosion of the blades which occurs when condensed water is blasted onto the blades (moisture carry over). Also, liquid water entering the blades may damage the thrust bearings for the turbine shaft. To prevent this, along with controls and baffles in the boilers to ensure high quality steam, condensate drains are installed in the steam piping leading to the turbine.

Maintenance requirements of modern steam turbines are simple and incur low costs (typically around $0.005 per kWh); their operational life often exceeds 50 years.

Speed Regulation

The control of a turbine with a governor is essential, as turbines need to be run up slowly to prevent damage and some applications (such as the generation of alternating current electricity) require precise speed control.

Uncontrolled acceleration of the turbine rotor can lead to an overspeed trip, which causes the nozzle valves that control the flow of steam to the turbine to close. If this fails then the turbine may continue accelerating until it breaks apart, often catastrophically. Turbines are expensive to make, requiring precision manufacture and special quality materials. During normal operation in synchronization with the electricity network, power plants are governed with a five percent droop speed control. This means the full load speed is 100% and the no-load speed is 105%. This is required for the stable operation of the network without hunting and drop-outs of power plants. Normally the changes in speed are minor. Adjustments in power output are made by slowly raising the droop curve by increasing the spring pressure on a centrifugal governor. Generally this is a basic system requirement for all power plants because the older and newer plants have to be compatible in response to the instantaneous changes in frequency without depending on outside communication.

Thermodynamics of Steam Turbines

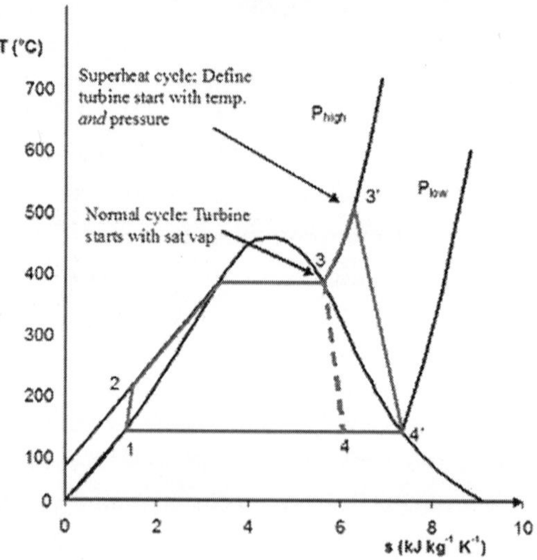

Fig. T-s diagram of a superheated Rankine cycle.

The steam turbine operates on basic principles of thermodynamics using the part 3-4 of the Rankine cycle shown in the adjoining diagram. Superheated steam (or dry saturated steam, depending on application) leaves the boiler at high temperature and high pressure. At entry to the turbine, the steam gains kinetic energy by passing through a nozzle (a fixed nozzle in an impulse type turbine or the fixed blades in a reaction type turbine). When the steam leaves the nozzle it is moving at high velocity towards the blades of the turbine rotor.

A force is created on the blades due to the pressure of the vapor on the blades causing them to move. A generator or other such device can be placed on the shaft, and the energy that was in the steam can now be stored and used. The steam leaves the turbine as a saturated vapor (or liquid-vapor mix depending on application) at a lower temperature and pressure than it entered with and is sent to the condenser to be cooled. The first law enables us to find an formula for the rate at which work is developed per unit mass. Assuming there is no heat transfer to the surrounding environment and that the changes in kinetic and potential energy are negligible compared to the change in specific enthalpy we arrive at the following equation

$$\frac{\dot{W}}{\dot{m}} = h_3 - h_4$$

where

- W Is the rate at which work is developed per unit time
- m A is the rate of mass flow through the turbine

Isentropic Efficiency

To measure how well a turbine is performing we can look at its isentropic efficiency. This compares the actual performance of the turbine with the performance that would be achieved by an ideal, isentropic, turbine. When calculating this efficiency, heat lost to the surroundings is assumed to be zero. The starting pressure and temperature is the same for both the actual and the ideal turbines, but at turbine exit the energy content ('specific enthalpy') for the actual turbine is greater than that for the ideal turbine because of irreversibility in the actual turbine. The specific enthalpy is evaluated at the same pressure for the actual and ideal turbines in order to give a good comparison between the two.

The isentropic efficiency is found by dividing the actual work by the ideal work.

$$\eta_t = \frac{h_3 - h_4}{h_3 - h_{4s}}$$

where

- h_3 is the specific enthalpy at state three
- h_4 is the specific enthalpy at state 4 for the actual turbine
- h_{4s} is the specific enthalpy at state 4s for the isentropic turbine

(but note that the adjacent diagram does not show state 4s: it is vertically below state 3)

DIRECT DRIVE

Electrical power stations use large steam turbines driving electric generators to produce most (about 80%) of the world's electricity. The advent of large steam turbines made central-station electricity generation practical, since reciprocating steam engines of large rating became very bulky, and operated at slow speeds.

Most central stations are fossil fuel power plants and nuclear power plants; some installations use geothermal steam, or use concentrated solar power (CSP) to create the steam.

Fig. A direct-drive 5 MW steam turbine fuelled with biomass.

Steam turbines can also be used directly to drive large centrifugal pumps, such as feedwater pumps at a thermal power plant. The turbines used for electric power generation are most often directly coupled to their generators. As the generators must rotate at constant synchronous speeds according to the frequency of the electric power system, the most common speeds are 3,000 RPM for 50 Hz systems, and 3,600 RPM for 60 Hz systems. Since nuclear reactors have lower temperature limits than fossil-fired plants, with lower steam quality, the turbine generator sets may be arranged to operate at half these speeds, but with four-pole generators, to reduce erosion of turbine blades.

MARINE PROPULSION

Fig. High and low pressure turbines forSS Maui.

Fig. Parsons turbine from the 1928 Polish destroyer Wicher.

In steamships, advantages of steam turbines over reciprocating engines are smaller size, lower maintenance, lighter weight, and lower vibration. A steam turbine is only efficient when operating in the thousands of RPM, while the most effective propeller designs are for speeds less than 300 RPM; consequently, precise (thus expensive) reduction gears are usually required, although numerous early ships through World War I, such as Turbinia, had direct drive from the steam turbines to the propeller shafts. Another alternative is turbo-electric transmission, in which an electrical generator run by the high-speed turbine is used to run one or more slow-speed electric motors connected to the propeller shafts; precision gear cutting may be a production bottleneck during wartime. Turbo-electric drive was most used in large US warships designed during World War I and in some fast liners, and was used in some troop transports and mass-production destroyer escorts in World War II.

The higher cost of turbines and the associated gears or generator/motor sets is offset by lower maintenance requirements and the smaller size of a turbine when compared to a reciprocating engine having an equivalent power, although the fuel costs are higher than a diesel engine because steam turbines have lower thermal efficiency.

To reduce fuel costs the thermal efficiency of both types of engine have been improved over the years. Today, propulsion steam turbine cycle efficiencies have yet to break 50%, yet diesel engines routinely exceed 50%, especially in marine applications. Diesel power plants also have lower operating costs since fewer operators are required. Thus, conventional steam power is used in very few new ships. An exception is LNG carriers which often find it more economical to use boil-off gas with a steam turbine than to re-liquify it. Nuclear-powered ships and submarines use a nuclear reactor to create steam for turbines.

Nuclear power is often chosen where diesel power would be impractical (as in submarine applications) or the logistics of refuelling pose significant problems (for example,icebreakers). It has been estimated that the reactor fuel for the Royal Navy's Vanguard-class submarines is sufficient to last 40 circumnavigations of the globe – potentially sufficient for the vessel's entire service life. Nuclear propulsion has only been applied to a very few commercial vessels due to the expense of maintenance and the regulatory controls required on nuclear systems and fuel cycles.

Early Development

The development of steam turbine marine propulsion from 1894-1935 was dominated by the need to reconcile the high efficient speed of the turbine with the low efficient speed (less than 300 rpm) of the ship's propeller at an overall cost competitive with reciprocating engines. In 1894, efficient reduction gears were not available for the high powers required by ships, so direct drive was necessary. InTurbinia, which has direct drive to each propeller shaft, the efficient speed of the turbine was reduced after initial trials by directing the steam flow through all three direct drive turbines (one on each shaft) in series, probably totaling around 200 turbine stages operating in series. Also, there were three propellers on each shaft for operation at high speeds.

The high shaft speeds of the era are represented by one of the first US turbine-powered destroyers, USS Smith, launched in 1909, which had direct drive turbines and whose three shafts turned at 724 rpm at 28.35 knots. The use of turbines in several casings exhausting steam to each other in series became standard in most subsequent marine propulsion applications, and is a form of cross-compounding.

The first turbine was called the high pressure (HP) turbine, the last turbine was the low pressure (LP) turbine, and any turbine in between was an intermediate pressure (IP) turbine. A much later arrangement than Turbinia can be seen on RMS Queen Mary in Long Beach, California, launched in 1934, in which each shaft is powered by four turbines in series connected to the ends of the two input shafts of a single-reduction gearbox. They are the HP, 1st IP, 2nd IP, and LP turbines.

Cruising Machinery and Gearing

The quest for economy was even more important when cruising speeds were considered. Cruising speed is roughly 50% of a warship's maximum speed and 20-25% of its maximum power level. This would be a speed used on long voyages when fuel economy is desired. Although this brought the propeller speeds down to an efficient range, turbine efficiency was greatly reduced, and early turbine ships had poor cruising ranges. A solution that proved useful through most of the steam turbine propulsion era was the cruising turbine. This was an extra turbine to add even more stages, at first attached directly to one or more shafts, exhausting to a stage partway along the HP turbine, and not used at high speeds.

As reduction gears became available around 1911, some ships, notably the battleship USS Nevada, had them on cruising turbines while retaining direct drive main turbines. Reduction gears allowed turbines to operate in their efficient range at a much higher speed than the shaft, but were expensive to manufacture.

Cruising turbines competed at first with reciprocating engines for fuel economy. An example of the retention of reciprocating engines on fast ships was the famous RMS Titanicof 1911, which along with her sisters RMS Olympic and HMHS Britannic had triple-expansion engines on the two outboard shafts, both exhausting to an LP turbine on the center shaft.

After adopting turbines with the Delaware-class battleships launched in 1909, the United States Navy reverted to reciprocating machinery on the New York-class battleships of 1912, then went back to turbines on Nevada in 1914. The lingering fondness for reciprocating machinery was because the US Navy had no plans for capital ships exceeding 21 knots until after World War I, so top speed was less important than economical cruising. The United States had acquired the Philippines and Hawaii as territories in 1898, and lacked the British Royal Navy's worldwide network of coaling stations.

Thus, the US Navy in 1900-1940 had the greatest need of any nation for fuel economy, especially as the prospect of war with Japan arose following World War I. This need was compounded by the US not launching any cruisers 1908-1920, so destroyers were required to perform long-range missions usually assigned to cruisers. So, various cruising solutions were fitted on US destroyers launched 1908-1916. These included small reciprocating engines and geared or ungeared cruising turbines on one or two shafts. However, once fully geared turbines proved economical in initial cost and fuel they were rapidly adopted, with cruising turbines also included on most ships. Beginning in 1915 all new Royal Navy destroyers had fully geared turbines, and the United States followed in 1917.

In the Royal Navy, speed was a priority until the Battle of Jutland in mid-1916 showed that in the battlecruisers too much armour had been sacrificed in its pursuit. The British used exclusively turbine-powered warships from 1906. Because they recognized that a significant cruising range would be desirable given their world-wide empire, some warships, notably the Queen Elizabeth-class battleships, were fitted with cruising turbines from 1912 onwards following earlier experimental installations.

In the US Navy, the Mahan-class destroyers, launched 1935-36, introduced double-reduction gearing. This further increased the turbine speed above the shaft speed, allowing smaller turbines than single-reduction gearing. Steam pressures and temperatures were also increasing progressively, from 300 psi/425 F (2.07 MPa/218 C)(saturation temperature) on the World War I-era Wickes class to 615 psi/850 F (4.25 MPa/454 C) superheated steam on some World War II Fletcher-class destroyers and later ships.A standard configuration emerged of an axial-flow high pressure turbine (sometimes with a cruising turbine attached) and a double-axial-flow low pressure turbine connected to a double-reduction gearbox.

This arrangement continued throughout the steam era in the US Navy and was also used in some Royal Navy designs. Machinery of this configuration can be seen on many preserved World War II-era warships in several countries. When US Navy warship construction resumed in the early 1950s, most surface combatants and aircraft carriers used 1,200 psi/950 F (8.28 MPa/510 C) steam. This continued until the end of the US Navy steam-powered warship era with the Knox-classfrigates of the early 1970s. Amphibious and auxiliary ships continued to use 600 psi (4.14 MPa) steam post-World War II, with USS Iwo Jima, launched in 2001, possibly the last non-nuclear steam-powered ship built for the US Navy.

Turbo-Electric Drive

Fig. NS 50 Let Pobedy, a nuclear icebreaker with nuclear-turbo-electric propulsion.

Turbo-electric drive was introduced on the battleship USS New Mexico, launched in 1917. Over the next eight years the US Navy launched five additional turbo-electric-powered battleships and two aircraft carriers (initially ordered as Lexington-class battlecruisers). Ten more turbo-electric capital ships were planned, but cancelled due to the limits imposed by the Washington Naval Treaty. AlthoughNew Mexico was refitted with geared turbines in a 1931-33 refit, the remaining turbo-electric ships retained the system throughout their careers.

This system used two large steam turbine generators to drive an electric motor on each of four shafts. The system was less costly initially than reduction gears and made the ships more maneuverable in port, with the shafts able to reverse rapidly and deliver more reverse power than with most geared systems. Some ocean liners were also built with turbo-electric drive, as were some troop transports and mass-production destroyer escorts in World War II. However, when the US designed the "treaty cruisers", beginning withUSS Pensacola launched in 1927, geared turbines were used for all fast steam-powered ships thereafter.

Current Usage

Since the 1980s, steam turbines have been replaced by gas turbines on fast ships and by diesel engines on other ships; exceptions are nuclear-powered ships and submarinesand LNG carriers. In the U.S. Navy, the conventionally powered steam turbine is still in use on all but one of the Wasp-class amphibious assault ships. The U.S. Navy also operates steam turbines on their nuclear powered Nimitz-class and Ford-class aircraft carriers along with all of their nuclear submarines (Ohio-,Los Angeles-, Seawolf-, andVirginia-classes). The Royal Navy decommissioned its last conventional steam-powered Leander-class frigate in 1993, also converting its sole Type 82 destroyer, HMS Bristol, into a training ship that same year. In 2013, the French Navy ended its steam era with the decommissioning of its last Tourville-class frigate.

Of the remaining blue-water navies, the Russian and Chinese navies currently operate steam-powered Kuznetsov-class aircraft carriers and Sovremenny-class destroyers, with China also operating steam-powered Luda-class destroyers. The Indian Navy currently operates two conventional steam-powered carriers, INS Viraat, a former British Centaur-class aircraft carrier (to be decommissioned in 2016), and INS Vikramaditya, a modified Kiev-class aircraft carrier; it also operates three Brahmaputra-class frigatescommissioned in the early 2000s and two Godavari-class frigates currently in the process of being decommissioned. The JDS Kurama, the last steam-powered JMSDF Shirane-class destroyer, will be decommissioned and replaced in 2017. Most other naval forces either retired or re-engined their steam-powered warships by 2010; as of 2015, the Brazilian Navy operates São Paulo, a former French Clemenceau-class aircraft carrier, while the Mexican Navy currently operates four former U.S. Knox-class frigates and two former U.S. Bronstein-class frigates. The Peruvian Navy currently operates the former Dutch De Zeven Provinciën-class cruiser BAP Almirante Grau; the Ecuadorian Navycurrently operates two modified Leander-class frigates.

LOCOMOTIVES

A steam turbine locomotive engine is a steam locomotive driven by a steam turbine. The main advantages of a steam turbine locomotive are better rotational balance and reduced hammer blow on the track. However, a disadvantage is less flexible output power so that turbine locomotives were best suited for long-haul operations at a constant output power.

The first steam turbine rail locomotive was built in 1908 for the Officine Meccaniche Miani Silvestri Grodona Comi, Milan, Italy. In 1924 Krupp built the steam turbine locomotive T18 001, operational in 1929, for Deutsche Reichsbahn.

TESTING

British, German, other national and international test codes are used to standardize the procedures and definitions used to test steam turbines. Selection of the test code to be used is an agreement between the purchaser and the manufacturer, and has some significance to the design of the turbine and associated systems.

In the United States, ASMEhas produced several performance test codes on steam turbines. These include ASME PTC 6-2004, Steam Turbines, ASME PTC 6.2-2011, Steam Turbines in Combined Cycles, PTC 6S-1988, Procedures for Routine Performance Test of Steam Turbines. These ASME performance test codes have gained international recognition and acceptance for testing steam turbines. The single most important and differentiating characteristic of ASME performance test codes, including PTC 6, is that the test uncertainty of the measurement indicates the quality of the test and is not to be used as a commercial tolerance.

EMERGENCY DIESEL GENERATOR

Emergency power requirements for the Patriot State are provided for by a 200 kW alternating current diesel driven generator. The engine is a Detriot Diesel Engine, model 8V71, and it is installed in the emergency diesel generating room, which is located on the bridge deck between frames 124 to 128, port side. It is located there because Coast Guard Regulations do not permit the emergency generator to be located within the main machinery spaces. Also, it is near the atmosphere for intake, exhaust, and ventilation purposes. The function of the emergency diesel generator is to supply power to the emergency switchboard in the event of a low voltage available at the main switchboard. A low voltage situation would occur when the bus voltage is 300 volts or less. The main switchboard would go into a low voltage situation if a casualty occured where the ship's service generators were unable to supply rated voltage to the main switchboard. The emergency diesel generator is capable of supplying all the loads fed by the emergency switchboard.

The emergency diesel generator is also used during cold ship starting and emergencies to supply limited power to engineroom machinery. If the main switchboard is deenergized, you can defeat the interlock on the bus tie breaker, and feed power back to the main switchboard. This will give you a very limited amount of power for loads such as the forced draft fan, fuel oil pump, etc. In addition, Machinery space lighting can be fed from the emergency switchboard without energizing any part of the main switchboard. The emergency diesel generator may not be paralleled with any of the ships service generators due to the interlock on the bus tie breaker.

The emergency generator will start automatically in the event of the a power failure. When the emergency diesel starts, it is important to check it soon after it has picked up the emergency load.

1. Check lube oil pressure, it should be 40 psig or higher.
2. Check cooling water temperature, it should not go above 190° F. Some of the load may have be to removed if it goes above 190° F.
3. Check the air louvres for proper opening by looking at their position. Check both inlet and outlet louvres.

The emergency alternating current diesel driven generator is rated at 200 kw, 450 volts, 0.8 power factor, 60 Hz, and it generates three phase electricity. The generator is designed to run at 1800 rpm The emergency diesel generator was not designed with an overload rating and must not be used outside the normal service ratings.

DIESEL GENERATOR SPECIFICATIONS

General Data	
Model number	70837305
Cycles (stroke)	2
Number of cylinders	8
Bore (in)	4.25
Stroke (in)	5
Total displacement	568
rpm 60 Hz at 1800	rpm
Compression ratio	17:1
Firing order	1L-3R-3L-4R-4L-2R-2L-1R
Starting system	Hydraulic
Crankshaft rotation	Right
Cylinder arrangement	V
Recommended Operating Conditions	
Lube oil pressure	50 - 70 psi
Fuel oil injection pressure	2200 - 3000 psi
Lube oil temperature	200 - 250°F.
Cooling water temperature	160 - 185°F. Do not exceed 210°F.
Fuel pump pressure	58 - 70 psi
Compression pressure	450 psi

FUEL OIL SERVICE SYSTEM

The fuel system consists of the fuel injectors, fuel pipes, fuel pump, fuel strainer, fuel filter and the necessary connecting lines.

A restricted fitting is located in the outlet passage in one of the cylinder heads to maintain pressure in the fuel system.

Fuel is drawn from the supply tank through the fuel strainer and enters the inlet to the fuel pump. The gear type fuel pump raises the fuel pressure and forces the fuel through the fuel filter. The fuel passes through the fuel filter into the fuel inlet manifold where it passes through fuel pipes into the inlet side of each fuel injector. The fuel is filtered through elements in the fuel injectors and atomized through small spray tip orifices into the combustion chamber. Surplus fuel, returning from the injectors, passes through the fuel return manifold and connecting fuel lines back to the fuel tank. The continuous flow of fuel through the injectors helps to cool the injectors and remove air from the fuel system.

FUEL INJECTOR

The fuel injector combines in a single unit, all of the parts necessary to provide complete and independent fuel injection at each cylinder. The injector creates the high pressure for fuel injection, meters the proper amount of fuel, atomizes the fuel, and time the injection into the combustion chamber.

FUEL OIL TRANSFER SYSTEM

The fuel oil transfer system originates in the lower engine room machinery spaces. A machinery space tank (port side D.B.) supplies fuel to the diesel oil day tank located on the flying bridge deck which holds 350 gallons and flows into a day tank in the emergency diesel room. The combination of the two tanks holds approximately 750-800 gallons of oil and is enough of a supply for 24 hours of diesel generator operation under full load.

Normal suction from the day tank is accomplished by the fuel oil booster pump which is engine driven, and mounted on the auxiliary drive housing. It draws fuel oil from the day tank through a Raycor suction filter and discharges it through the pressure filter to the fuel oil header. From this header, fuel oil flows to the individual fuel injection pumps which deliver the fuel oil to the injection nozzles in the proper quantity. There is a pressure relief valve which is between the transfer pump and the pressure filter which is set to open at 100 psi. When this valve opens, the fuel oil is by-passed back to the fuel oil day tank. A pressure relief valve connected to the fuel oil header is set to open at a pressure of 35 pounds. When this valve opens, the fuel oil is by-passed back to the fuel oil tank. The lighter the load on the engine, the greater will be the quantity of oil passing through this valve, thus maintaining a constant head of fuel oil.

COOLING WATER SYSTEM

The cooling system is of the closed type with the cooling water flowing from a radiator holding approximately 8.8 gallons. Coolant is drawn from the lower portion of the radiator by the water pump and is forced through the oil cooler housing into the cylinder block. From the cylinder block, the coolant passes up through the cylinder heads and when the engine is at normal operating temperature, through the thermostats and into the upper portion of the radiator. The coolant passed down a series of tubes where its temperature is lowered by the air stream created by the revolving fan. A centrifugal pump driven from the engine camshaft at the pump end of the engine circulates the water through the system.

LUBRICATION

The oil pump on the 8V engines is placed in the crankshaft front cover and consists of a pair of spur gears, one large and one small, which mesh together and ride in a cavity inside the crankshaft cover. The larger gear is concentric with and splined to a pump drive hub on the front end of the crankshaft. The pump idler gear is much smaller and runs on a bushing and hardened steel shaft pressed into the crankshaft cover. The oil is drawn by suction from the oil pan through the intake screen and pipe to the oil pump where it is pressurized. The oil then passes from the pump into a short gallery in the cylinder block to the oil cooler adaptor plate. At the same time, oil from the pump is directed to a spring loaded pressure relief valve mounted on the cylinder block. This valve discharges excess oil directly to the oil sump when the pump pressure exceeds 100 psi (689 KPa).

From the main oil gallery, the pressurized oil flows through drilled passages to each main bearing then passes to an adjacent pair of connecting rods by means of grooves in the unloaded halves of the main and connecting rod bearings and drilled passages in the crankshaft. The rifle drilled connecting rods carry oil from the rod bearings to the piston pin bushings and to the nozzles at the upper end of each connecting rod which provides the cooling oil spray for the piston crown. At the rear of the block, two diagonally drilled oil passages which intersect the main oil gallery carry oil to two rear camshaft end bearings. Oil is then conducted through the rifle drilled camshaft to the intermediate and front end bearings. Oil from the camshaft intermediate bearing is directed against the camshaft lobes and cam rollers which run in an oil bath. This oil from the intermediate bearing provides lubrication of the cam lobes immediately after starting the engine when the oil is cold and before camshaft bearing oil flow and oil drainage from the cylinder head have had time to build up.

The diagonally drilled oil passage on the right side of the rear of the block intersects with a vertical passage to carry oil to the right bank cylinder head. A short gallery also intersects with the diagonal passage to lubricate the idler gear bearings. Another gallery intersecting the diagonal passages at the front of the block supplies oil to the left bank cylinder head. Drilled passages, intersecting longitudinal galleries which parallel the camshaft, lead to blowers and supply oil for the blower drive gears and bearings.

CRANKCASE VENTILATION SYSTEM

A gear driven crankcase exhauster is provided to pull a slight vacuum on the crankcase at all times. This provision prevents the formation of dangerous pockets of gas accumulation in the crankcase and assures that gasket leakage will be maintained at a minimum for all crank pit covers and cylinder block closures. Spring Loaded explosion covers, quick opening and quick closing, are provided at each end of the engine. These covers are of sufficient area to relieve pressure above 13 psi which may build up in the crankcase due to mechanical troubles or explosions due to local hot spots.

GOVERNOR AND SPEED CONTROL

Engine speed is maintained by an Woodward governor which is hydraulic and of the isochronous type. An isochronous governor will maintain the desired engine speed regardless of load. A hand speed control lever is furnished to control speeds between idle and 1800 rpm (full speed). Load should be applied only when the throttle is up all the way. The engine is also equipped with a mechanical overspeed trip in case of governor failure.

The movements of the governor power mechanism are transmitted through the lever and link connections to the fuel pump control rod. These connections are fitted with hardened pins and needle bearings which require oil periodically. A light spring fastened to the control rod takes up lost motion due to linkage wear. A bell crank and block located near the governor can be adjusted to give proper rack travel of 18mm maximum. A compression link assembly is provided between the fuel pump control rod governor, to disconnect the governor from the fuel pump control rod when the overspeed trip is tripped. When the overspeed trip is reset, the compression link assembly returns to its normal position.

HYDRAULIC STARTING SYSTEM

Engine starting is acomplished by a hydraulic starting system. Hydraulic fluid is stored in four nitrogen charged accumulators at a pressure of 2800 psi. To start the engine, a hydraulic control valve is pulled forward, hydraulic fluid is released through a high pressure hose to the starter mounted on the left side of the Detroit Diesel. The starter motor turns over the engine, and the hydraulic fluid returns to the hydraulic resevoir tank. The accumulators can be recharged by an electric motor or by a hand crank when electrical power is not present.

PREVENTIVE MAINTENANCE

In order to insure proper operation of the engine at all times, preventive maintenance is an absolute necessity. When not being used under normal operation, this engine should be run for at least a half hour once a week. Lube oil filter should be changed every 30 hours of operation. Along with this, lube oil filters and fuel oil filters should be changed when they become heavy with sludge and dirt particle. Always check that the crankcase contains oil up to the operating level, and that there is a sufficient amount of fresh water in the engine for proper cooling before starting up.

Always keep a sufficient level of fuel oil in the day tank during engine operation (at least half full). Maintaining at least that level ensures against stirring up dirt and sludge that may be at the bottom of the tank.

DIESEL GENERATOR

Fig. A Cummins diesel generator of 500kVA temporarily parked in a tourist resort in Egypt.

Fig. A 200 kW Caterpillar diesel generator set in a sound attenuated enclosure used as emergency backup at a sewage treatment substation inAtlanta, United States.

A diesel generator is the combination of a diesel engine with an electric generator (often an alternator) to generate electrical energy. This is a specific case of engine-generator. A diesel compression-ignition engine often is designed to run on fuel oil, but some types are adapted for other liquid fuels or natural gas. Diesel generating sets are used in places without connection to a power grid, or as emergency power-supply if the grid fails, as well as for more complex applications such as peak-lopping, grid support and export to the power grid.

Sizing of diesel generators is critical to avoid low-load or a shortage of power and is complicated by modern electronics, specifically non-linear loads. In size ranges around 50 MW and above, an open cycle gas turbine is more efficient at full load than an array of diesel engines, and far more compact, with comparable capital costs; but for regular part-loading, even at these power levels, diesel arrays are sometimes preferred to open cycle gas turbines, due to their superior efficiencies.

DIESEL GENERATOR SET

Fig. Diesel generator on an oil tanker.

The packaged combination of a diesel engine, a generator and various ancillary devices (such as base, canopy, sound attenuation,control systems, circuit breakers, jacket water heaters and starting system) is referred to as a "generating set" or a "genset" for short.

Set sizes range from 8 to 30 kW (also 8 to 30 kVA single phase) for homes, small shops and offices with the larger industrial generators from 8 kW (11 kVA) up to 2,000 kW (2,500 kVA three phase) used for large office complexes, factories. A 2,000 kW set can be housed in a 40 ft (12 m) ISO container with fuel tank, controls, power distribution equipment and all other equipment needed to operate as a standalone power station or as a standby backup to grid power. These units, referred to as power modules are gensets on large triple axle trailers weighing 85,000 pounds (38,555 kg) or more.

A combination of these modules are used for small power stations and these may use from one to 20 units per power section and these sections can be combined to involve hundreds of power modules. In these larger sizes the power module (engine and generator) are brought to site on trailers separately and are connected together with large cables and a control cable to form a complete synchronized power plant. A number of options also exist to tailor specific needs, including control panels for autostart and mains paralleling, acoustic canopies for fixed or mobile applications, ventilation equipment, fuel supply systems, exhaust systems, etc. Diesel generators, sometimes as small as 200 kW (250 kVA) are widely used not only for emergency power, but also many have a secondary function of feeding power to utility grids either during peak periods, or periods when there is a shortage of large power generators.

Ships often also employ diesel generators, sometimes not only to provide auxiliary power for lights, fans, winches etc., but also indirectly for main propulsion. With electric propulsion the generators can be placed in a convenient position, to allow more cargo to be carried. Electric drives for ships were developed prior to World War I. Electric drives were specified in many warships built during World War II because manufacturing capacity for large reduction gears was in short supply, compared to capacity for manufacture of electrical equipment. Such a diesel-electric arrangement is also used in some very large land vehicles such as railroad locomotives.

GENERATOR SIZE

Generating sets are selected based on the electrical load they are intended to supply, the electrical load's characteristics such as kWe, kVA, var, harmonic content, surge currents (e.g., motor starting current) and non-linear loads. The expected duty (such as emergency, prime or continuous power) as well as environmental conditions (such asaltitude, temperature and exhaust emissions regulations) must also be considered. Most of the larger generator set manufacturers offer software that will perform the complicated sizing calculations by simply inputting site conditions and connected electricalload characteristics.

POWER PLANTS – ELECTRICAL "ISLAND" MODE

One or more diesel generators operating without a connection to an electrical grid are referred to as operating in island mode. Operating generators in parallel provides the advantage of redundancy, and can provide better efficiency at partial loads. The plant brings generator sets online and takes them off line depending on the demands of the system at a given time. An islanded power plant intended for primary power source of an isolated community will often have at least three diesel generators, any two of which are rated to carry the required load. Groups of up to 20 are not uncommon.

Generators can be electrically connected together through the process of synchronization. Synchronization involves matching voltage, frequency and phase before connecting the generator to the system. Failure to synchronize before connection could cause a high short circuit current or wear and tear on the generator or its switchgear. The synchronization process can be done automatically by an auto-synchronizer module, or manually by the instructed operator. The auto-synchronizer will read the voltage, frequency and phase parameters from the generator and busbar voltages, while regulating the speed through the engine governor or ECM (Engine Control Module).

Load can be shared among parallel running generators through load sharing. Load sharing can be achieved by using droop speed control controlled by the frequency at the generator, while it constantly adjusts the engine fuel control to shift load to and from the remaining power sources. A diesel generator will take more load when the fuel supply to its combustion system is increased, while load is released if fuel supply is decreased.

SUPPORTING MAIN UTILITY GRIDS

In addition to their well known role as power supplies during power failures, diesel generator sets also routinely support main power grids worldwide in two distinct ways:

Grid Support

Emergency standby diesel generators, for example such as those used in hospitals, water plant, are, as a secondary function, widely used in the US and, in the recent past, in Great Britain (Short Term Operating Reserve) to support the respective national grids at times for a variety of reasons. The tenders known as the Short Term Operating Reservehave exhibited quite variable prices, and from 2012 the volume of demand-side participation, which mainly entails the use of on-site diesels, has dropped as the tendered prices fell. In the UK for example, some 0.5 GWe of diesels have at times been used to support the National Grid, whose peak load is about 60 GW. These are sets in the size range 200 kW to 2 MW. This usually occurs during, for example, the sudden loss of a large conventional 660 MW plant, or a sudden unexpected rise in power demand eroding the normal spinning reserve available. This is beneficial for both parties - the diesels have already been purchased for other reasons; but to be reliable need to be fully load tested. Grid paralleling is a convenient way of doing this. This method of operation is normally undertaken by a third party aggregator who manages the operation of the generators and the interaction with the system operator.

These diesels can in some cases be up and running in parallel as quickly as two minutes, with no impact on the site (the office or factory need not shut down). This is far quicker than a base load power station which can take 12 hours from cold, and faster than a gas turbine, which can take several minutes. Whilst diesels are very expensive in fuel terms, they are only used a few hundred hours per year in this duty, and their availability can prevent the need for base load station running inefficiently at part load continuously. The diesel fuel used is fuel that would have been used in testing anyway.

In Great Britain, National Grid can generally rely upon about 2 GW of customer demand reduction via back-up diesels being self-despatched for about 10 to 40 hours a year at times of expected peak national demand. National Grid does not control these diesels - they are run by the customer to avoid "triad" transmission network use of system (TNUoS) charges which are levied only on consumption of each site, at the three half-hours of peak national demand. It is not known in advance when the three half-hours of peak national demand (the "triad" periods) will be, so the customer must run his diesels for a good deal more half-hours a year than just three.

The total capacity of reliably operable standby generation in Britain is estimated to be around 20 GW, nearly all of which is driven by diesel engines. This is equivalent to nearly 29% of the British system peak, although only a very small fraction will ever be generating at the same time. Most plant is for large offices blocks, hospitals, supermarkets, and various installations where continuous power is important such as airports. Therefore most is in urban areas, particularly city and commercial centres. It is estimated that around 10% of plant exceeds 1 MW, about 50% is in the 200 kW-1 MW range, and the remaining 40% is sub-200 kW. Although it is growing, only a very small proportion is believed to be used regularly for peak lopping, the vast majority just being only for standby generation. The information in this paragraph is sourced from section 6.9 of the government report: "Overcoming Barriers To Scheduling Embedded Generation To Support Distribution Networks"

A similar system to Great Britain's (Short Term Operating Reserve), operates in France. It is known as EJP; at times of grid stress, special tariffs can mobilize at least 5 GW of diesel generating sets to become available. In this case, the diesels prime function is to feed power into the grid.

During normal operation in synchronization with the electricity net, powerplants are governed with a five percent droop speed control. This means the full load speed is 100% and the no load speed is 105%. This is required for the stable operation of the net without hunting and dropouts of power plants. Normally the changes in speed are minor. Adjustments in power output are made by slowly raising the droop curve by increasing the spring pressure on a centrifugal governor. Generally this is a basic system requirement for all powerplants because the older and newer plants have to be compatible in response to the instantaneous changes in frequency without depending on outside communication.

COST OF GENERATING ELECTRICITY

Typical Operating Costs

Fuel consumption is the major portion of diesel plant owning and operating cost for power applications, whereas capital cost is the primary concern for backup generators. Specific consumption varies, but a modern diesel plant will at its near-optimal 65-70% loading, generate 3 kWh per litre.

GENERATOR SIZING AND RATING

Rating

Generators must provide the anticipated power required reliably and without damage and this is achieved by the manufacturer giving one or more ratings to a specific generator set model. A specific model of a generator operated as a standby generator may only need to operate for a few hours per year, but the same model operated as a prime power generator must operate continuously. When running, the standby generator may be operated with a specified - e.g. 10% overload that can be tolerated for the expected short running time. The same model generator will carry a higher rating for standby service than it will for continuous duty. Manufacturers give each set a rating based on internationally agreed definitions.

These standard rating definitions are designed to allow correct machine selection and valid comparisons between manufacturers to prevent them from misstating the performance of their machines, and to guide designers.

Generator Rating Definitions

Standby Rating based on Applicable for supplying emergency power for the duration of normal power interruption. No sustained overload capability is available for this rating. (Equivalent to Fuel Stop Power in accordance with ISO3046, AS2789, DIN6271 and BS5514). Nominally rated.

Typical application - emergency power plant in hospitals, offices, factories etc. Not connected to grid.

Prime (Unlimited Running Time) Rating: Should not be used for Construction Power applications. Output available with varying load for an unlimited time. Typical peak demand 100% of prime-rated ekW with 10% of overload capability for emergency use for a maximum of 1 hour in 12. A 10% overload capability is available for limited time. (Equivalent to Prime Power in accordance with ISO8528 and Overload Power in accordance with ISO3046, AS2789, DIN6271, and BS5514). This rating is not applicable to all generator set models. Typical application - where the generator is the sole source of power for say a remote mining or construction site, fairground, festival etc.

Base Load (Continuous) Rating based on: Applicable for supplying power continuously to a constant load up to the full output rating for unlimited hours. No sustained overload capability is available for this rating. Consult authorized distributor for rating. (Equivalent to Continuous Power in accordance with ISO8528, ISO3046, AS2789, DIN6271, and BS5514). This rating is not applicable to all generator set models Typical application - a generator running a continuous unvarying load, or paralleled with the mains and continuously feeding power at the maximum permissible level 8,760 hours per year. This also applies to sets used for peak shaving/grid support even though this may only occur for say 200 hours per year. As an example if in a particular set the Standby Rating were 1000 kW, then a Prime Power rating might be 850 kW, and the Continuous Rating 800 kW. However these ratings vary according to manufacturer and should be taken from the manufacturer's data sheet. Often a set might be given all three ratings stamped on the data plate, but sometimes it may have only a standby rating, or only a prime rating.

Sizing

Typically however it is the size of the maximum load that has to be connected and the acceptable maximum voltage drop which determines the set size, not the ratings themselves. If the set is required to start motors, then the set will have to be at least three times the largest motor, which is normally started first. This means it will be unlikely to operate at anywhere near the ratings of the chosen set. Many gen-set manufacturers have software programs that enable the correct choice of set for any given load combination. Sizing is based on site conditions and the type of appliances, equipment, and devices that will be powered by the generator set.

INSTALLATION

Manufacturers provide detailed installation guidelines to ensure correct functioning, reliability and low maintenance costs. There are a range of different styles of generators for camping, RV's standby power, and portable commercial applications. When purchasing a generator online you must ensure that the product comes with installation guidelines and that ts stipulated within the product description.

Guidelines cover such things as:

- Sizing and selection - typical loads of electrical devices, expected load for the application

- Electrical factors - Peak loads versus steady and allowing for certain applicances
- Cooling - Engine cooling during the power generation process
- Ventilation - Allowing air flow during the cooling process (design)
- Fuel storage - How long will the generator run for at a specific usage rate
- Noise - dB levels to be expected for specific models
- Exhaust - ppm(parts per million) levels of certain element, consider the environment and those around the generator
- Starting systems - key start, pull start, button start - lots of options.

ENGINE DAMAGE

Diesel engines can suffer damage as a result of misapplication or misuse - namely internal glazing (occasionally referred to as bore glazing or piling) and carbon build-up. Ideally, diesel engines should be run at least 60% to 75% of their maximum rated load. Short periods of low load running are permissible providing the set is brought up to full load, or close to full load on a regular basis. Internal glazing and carbon build-up is due to prolonged periods of running at low speeds or low loads. Such conditions may occur when an engine is left idling as a 'standby' generating unit, ready to run up when needed, (misuse); if the engine powering the set is over-powered (misapplication) for the load applied to it, causing the diesel unit to be under-loaded, or as is very often the case, when sets are started and run off load as a test (misuse).

Running an engine under low loads causes low cylinder pressures and consequent poor piston ring sealing since this relies on the gas pressure to force them against the oil film on the bores to form the seal. Low cylinder pressures causes poor combustion and resultant low combustion pressures and temperatures. This poor combustion leads to soot formation and unburnt fuel residues which clogs and gums piston rings, causing a further drop in sealing efficiency and exacerbates the initial low pressure. Glazing occurs when hot combustion gases blow past the now poorly-sealing piston rings, causing the lubricating oil on the cylinder walls to 'flash burn', creating an enamel-like glaze which smooths the bore and removes the effect of the intricate pattern of honing marks machined into the bore surface which are there to hold oil and return it to the crankcase via the scraper ring.

Hard carbon also forms from poor combustion and this is highly abrasive and scrapes the honing marks on the bores leading to bore polishing, which then leads to increased oil consumption (blue smoking) and yet further loss of pressure, since the oil film trapped in the honing marks is intended to maintain the piston seal and pressures. Unburnt fuel then leaks past the piston rings and contaminates the lubricating oil. Poor combustion causes the injectors to become clogged with soot, causing further deterioration in combustion and black smoking. The problem is increased further with the formation of acids in the engine oil caused by condensed water and combustion by-products which would normally boil off at higher temperatures. This acidic build-up in the lubricating oil causes slow but ultimately damaging wear to bearing surfaces. This cycle of degradation means that the engine soon becomes irreversibly damaged and may not start at all and will no longer be able to reach full power when required.

Under-loaded running inevitably causes not only white smoke from unburnt fuel but over time will be joined by blue smoke of burnt lubricating oil leaking past the damaged piston rings, and black smoke caused by damaged injectors. This pollution is unacceptable to the authorities and neighbors. Once glazing or carbon build up has occurred, it can only be cured by stripping down the engine and re-boring the cylinder bores, machining new honing marks and stripping, cleaning and de-coking combustion chambers, fuel injector nozzles and valves.

If detected in the early stages, running an engine at maximum load to raise the internal pressures and temperatures allows the piston rings to scrape glaze off the bores and allows carbon build-up to be burnt off. However, if glazing has progressed to the stage where the piston rings have seized into their grooves, this will not have any effect. The situation can be prevented by carefully selecting the generator set in accordance with manufacturers printed guidelines. (the use off additional depth oil and fuel By Pass filtration, down to <3 micron level can prevent buildup of the particulate or carbon build that contributes to the varnishing.)

For emergency only sets, it may be impractical to use the supported load for testing. A temporary or permanent load bank can be used testing. Sometimes the switchgear can be designed to allow the set to feed power into the grid for load testing.

FUELS

Diesel fuel is named after diesel engines, and not vice versa; diesel engines are simply compression-ignition engines, and can operate on a variety of different fuels, depending on configuration and location. Where a gas grid connection is available, gas is often used, as the gas grid will always remain pressurised even during almost all power cuts; in more rural situations, or for low load factor plant, diesel fuel derived from crude oil is a common fuel; it is less likely to freeze than heavier oils. Endurance will be limited by tank size.

Diesel engines can work with the full spectrum of crude oil distillates, from natural gas, alcohols, gasoline, wood gas to the fuel oils from diesel oil to cheaper residual fuels that are like lard at room temperature, and must be heated to enable them to flow down a fuel line. This is implemented by introducing gas with the intake air and using a small amount of diesel fuel for ignition. Conversion to 100% diesel fuel operation can be achieved instantaneously. Larger engines (from about 3 MWe to 30 MWe) sometimes use heavy oils, essentially tars, derived from the end of the refining process. The slight added complexity of keeping the fuel oil heated to enable it to flow, whilst mitigating the fire risks that come from over-heating fuel, make these fuels unpopular for smaller, often unmanned, generating stations.

4

Steam Generators

STEAM GENERATOR (BOILER)

A steam generator is a form of low water-content boiler, similar to a flash steam boiler. The usual construction is as a spiral coil of water-tube, arranged as a single, or monotube, coil. Circulation is once-through and pumped under pressure, as a forced-circulation boiler. The narrow-tube construction, without any large-diameter drums or tanks, means that they are safe from the effects of explosion, even if worked at high pressures. The pump flowrate is adjustable, according to the quantity of steam required at that time. The burner output is throttled to maintain a constant working temperature. The burner output required varies according to the quantity of water being evaporated: this can be either adjusted byopen-loop control according to the pump throughput, or by a closed-loop control to maintain the measured temperature.

They are used as auxiliary boilers on ships.

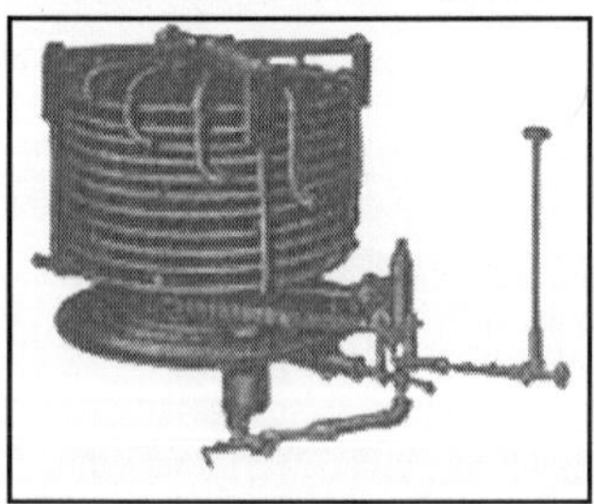

Fig. A monotube boiler, a type of steam generator (early 1900s White Motor Company).

TYPES

Stone-Vapor

One of the best-known designs is the Stone-Vapor. The inner casing of the boiler forms a vertical bell, with an outer airtight cylindrical casing. The oil or gas burner is mounted at the top, above the coils, and facing downwards.

The heating element is a single tube, arranged into a number of helical cylinders. The first helices (in the flow direction) are small-diameter tubes, wrapped in large diameter turns. Succeeding turns are coiled inside this and the tube is of progressively increasing diameter, to allow for a constant flow rate as the water evaporates into steam and forms bubbles. The steam outlet is from the final turn at the bottom of the inner helix. The outlet is approximately 90% steam (by mass) and residual water is separated by passing it through a steam-water separator. The exhaust gases turn upwards and flow over the outside of the bell, usually passing additional helices that are used as an initial feedwater heater.

Clayton

The Clayton steam generator is similar to the Stone-Vapor, but the burner and flow directions are reversed. The heating coil is mounted within a simple cylindrical casing. Rather than helical, cylindrical layers, the Clayton coils are arranged as layers of flat spirals. Water is pumped into the top layers and forced downwards. Again, the tube diameter increases in steps, as evaporation takes place. The final turns form a single closely spaced helical cylinder around the burner as a water-wall furnace and is heated by radiant heat. The steam output is passed through a centrifugal separator and a dry steam quality of 99.5% is claimed.

STEAM GENERATOR (AUXILIARY BOILER)

A steam generator, in the sense of an auxiliary boiler on a ship, is a form of boiler. It uses high-pressure superheated steam from the main steam supply used for propulsion, and uses this to generate low pressure saturated steam. This is then used to power auxiliary steam engines around the ship such as winches or pumps, or to supply any steam requirement on the ship that does not require high superheating, such as boiler feedwater and freshwater evaporators.

TYPES

Low-Pressure Steam Generators

A typical generator produces around 10,000 lb/hour at around 120 psi, using inlet steam supplied at 600 psi. They are constructed as a cylindrical steel pressure vessel, either horizontal or vertical, containing a series of heating coils. There is a strong resemblance to the design of evaporators, although steam generators can be rather simpler as they use treated feedwater, rather than scale-producing seawater.

STEAM BOILERS DEVELOPMENT

Steam boilers are not what they once were; Scotch fire tube boilers. The steam boilersdesign remained much the same and its qualities of ruggedness, reliability, ease of maintenance, and ability to stand abuse made it a very popular steam boiler in the marinefield. However there was a realization that advanced steam boiler designs with greater generating capacity, higher pressure, and increased efficiency were needed to meet the changing requirements brought about by the development of turbine propulsion machinery. The marine engineers then began to adapt various types of water tube steam boilers to fit into the restricted space available for installations on board steamships. It was also realized thatmarine boilers would have to be designed to burn efficiently whatever fuel oils would be available.

Although the steam boilers are basically the same, water boiling cauldrons, modern steam boilers are different in details, especially concerning the control systems. Marine steam boilers today are provided with automatic surveillance and they operate without supervision 24 hour per day. Just as the need for increased steaming capacities within fixed space and weight limitations spurred the development of the earlier water-tube steam boiler, so did the same need precipitate the advance to boiler designs of the bent-tube type.

The term bent-tube boiler covers a wide range of boiler designs characterized by the ability to raise steam pressure within a short period of time. This means smaller tube diameters, fewer headers, increased furnace ratings and wider application of water-cooled furnaces. The boilers of this category also became known as multi-drum boiler type, express boiler type, or by various other names corresponding to the designers of particular arrangements. With advanced designs available for naval steamships, it was quite logical to utilize adoptions of those designs for high powered passenger steamships.

STEAM GENERATORS AND BOILERS – SYSTEMS

Systems and equipment that ensure trouble-free operation of the steam generating plant are designed with the same Clayton experience and expertise. Clayton offers a range of ancillaries and packages to match the selected steam generators and manufactured to the same high standards.

The most common system is the Clayton Open System in which condensate can be returned at atmospheric pressure. In the Open System, boiler feedwater is directed from a Hotwell to the Clayton Pump and is then fed to the Clayton Steam Generator. Steam flows from the Clayton Steam Separator outlet nozzle and the condensate from the separator is re-cycled back to the hotwell. The small amount of blowdown required on a Clayton Steam Generator is taken from this point. The make-up water supply can be passed through a water softener and chemicals can be dosed directly into the hotwell.

On multiple Clayton Steam Generator installations the feedwater equipment is common to all. Other tried and tested designs include the Clayton Closed and Semi-Closed Systems that are used where energy savings from condensate heat recovery is required. High Pressure and Superheated Steam is another application where the design of the Clayton Steam Generator is perfectly suited. The small diameter water tube coil allows steam pressures of up to 200 barg to be produced and the helical heat exchanger concept means that superheated steam can be generated up to temperatures of 460 degrees centigrade.

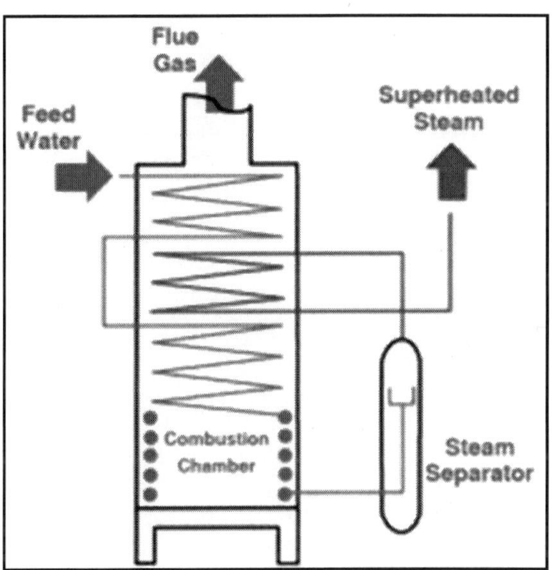

The spiral construction of the Clayton heating coil means that it is extremely robust and will counteract any effects of expansion and contraction during rapid heating and cooling cycles without experiencing any mechanical damage. On Superheated Clayton Steam Generators the feed water is directed to the topmost layer of coil where it is pre-heated as it spirals downwards and fed to the lower section from where it passes to the high efficiency steam separator. Dry steam from the separator then trevels to the integral superheating coil section where the desired steam temperature is produced.

Power Stations have benefited by using the Clayton Steam Generator where superheated steam, high pressure steam and rapid start capability combined with the inherent safety have particular benefits. In both conventional and nuclear power stations the Clayton Steam Generator is used for peak shaving, auxiliary start-up steam, back-up steam supply and emergency steam. When used as an auxiliary start-up boiler the Clayton Steam Generator enables a power plant to be brought on-line rapidly which means lower emissions, less fuel consumption, less water consumption, less chemical consumption, less electricity consumption and provides increased revenue as well as the commercial advantage that comes from having the flexibility of being able to start-up quickly.

Marine & Offshore Steam Systems are ideal applications where the safety, space saving, unmanned operation, reliability and modular construction have made Clayton the offshore standard for ships and offshore oil platforms. The Clayton Marine Steam Generator is type approved by many marine authorities and can be supplied to the requirements of any inspection agency.

Clayton's own marine standard includes specified materials, control systems, electrical motors, cabling, tagging, documentation and other essential features required by this demanding industry.

BOILERS

On the Patriot State, steam is produced for propulsion and other requirements by two Babcock and Wilcox two-drum, oil fired marine water tube boilers, that are capable of producing 108,000 lbs/hr of superheated steam at 620 psi and 905° F. Each boiler is fitted with four registers and steam atomized burners, soot blowers, smoke indicators, water gauges, high and low water alarms, and safety valves. Originally the boilers were fitted with control desuperheaters to maintain design temperature at the superheater outlet under varying operating conditions. The control desuperheaters were removed when the ship was converted to the training ship, and consequently, we have to carefully watch the superheater outlet temperature during high steaming conditions.

Furnace

The boiler furnace is made up of tubes covering four walls and a floor. The side wall bends upward toward the steam drum and therefore also form the furnace roof. The side, front and rear walls are water cooled with bare tube construction.

The floor is water cooled by tubes installed below two courses of firebrick. The front wall consists of 2" O.D. bare tubes expanded and belled into the upper and lower front wall headers. The lower front wall header is supplied with water by the 2" O.D. floor tubes, which connect with the water drum. Two 4-1/2" O.D. riser tubes connect the upper front wall header with the steam drum. The side wall and roof consist of 2" O.D. bare tubes expanded and belled into the sidewall header and the steam drum. The side wall header is supplied with water by the upper row of 2" O.D. floor tubes, which connect with the water drum. The rear wall consist of 2" O.D. bare tubes expanded and belled into the upper and lower rear wall headers. The lower rear wall header is supplied with water by the lower row of 2" O.D. floor tubes, which connect with the water drum. Three 4-1/2" O.D. riser tubes connect the upper rear wall header with the steam drum. The floor tubes consist of two rows of 2" O.D. tubes beneath the floor firebrick. They supply water from the water drum to the side and rear wall headers. These arrangements give complete side, front, rear and floor tube circulation independent of the main circulation circuit in the boiler generating tubes.

Fig. Two-drum, Single-furnace Boiler with horizontal superheater.

Steam Drum

A standard Babcock and Wilcox triple-plate perforated baffle is installed in the steam drum. Its purpose is to aid in the delivery of clean dry steam to the superheater via the dry pipe. This will aid in avoiding the deposit of solids on the surfaces exposed to the steam. The baffle plates themselves extend longitudinally along the interior of the drum and allows for adequate areas for the flow of released steam. The chemical feed pipe is used to inject boiler chemicals into the steam drum while the boiler is in operation, and also as a sampling pipe for testing the boiler water. The chemical feed pipe extends almost the entire length of the drum below the normal water level and connects to the chemical feed piping nozzle on the end of the drum. A row of 1/8" diameter holes are drilled along the top centerline of the pipe to pass the chemicals. There is a separate boiler water sampling line which can also be used when putting the boiler on continuous blow. The continuous blow is to the bilge. There is no surface blow line on these boilers.

The dry pipe is suspended along the top centerline inside the steam drum. It is perforated along its upper surface with rows of 3/8" diameter holes and drain holes are provided in the bottom of the pipe at each end. Steam enters the dry pipe through the upper perforated area and leaves the pipe through the connection which leaves from the top of the drum to the superheater. The dry pipe serves a dual function; it reduces moisture carry-over and provides for an even withdrawal of steam from the water surface. The internal feedline is that part of the feedline within the steam drum. It extends approximately 90% the length of the drum, below the normal water level, and is connected to the feed supply nozzle. It has one row of 1/2" diameter holes along the pipe to introduce the feedwater into the steam drum evenly throughout to prevent thermal shock.

Superheater

The superheater is made up of a series of 1-1/4" O.D. U-bent tubes arranged in four loops and connecting two superheater headers which are supported at the rear of the boiler. The U-tubes extend the full depth of the boiler and are supported by four 3/4" O.D. support tubes. The steam will make six passes of the superheater.

The boiler proper consist of a 54" steam drum and a 30" water drum connected by a generating bank of 18 rows of 1-1/4" O.D. tubes bent at the ends so they enter the boiler drums radially for expanding and belling. The bank is arranged so that the furnace gases make a single pass over it before entering the uptakes. Also, there are two rows of 2" O.D. tubes between the furnace and the superheater which screen the superheater from the direct radiation of the furnace gases. Safety valves are installed on both boilers to prevent the pressure in the boiler from rising above a safe working pressure. To insure positive flow through the superheater, the superheater safety valves will pop before the two drum safety valves. The safety valves from both boilers connect to a common escape pipe. A superheater outlet safety valve is installed at the superheater outlet and is the one that will lift when the first drum valve pops - and remain open until the last drum safety is seated. The safety valves on a boiler must be of sufficient size and number to reduce the steam drum pressure to a safe level at full firing rate with the boiler completely cutoff steam lines. % Blowdown = F((lifting pressure - reseating pressure) x 100,(working pressure)) To insure the valve is free, each safety valve or safety relief valve has a lifting device so the valve disk may be lifted from its seat by hand. There must be at least 75% full working pressure on the boiler before the hand easing gear is used.

Safety valve must never be LOCKED or GAGED other than when conducting a hydrostatic pressure test. The safety and relief valve escape system consists of two escape pipes led from the machinery spaces through the inside of the outer stack to the atmosphere. Eight inch escape piping is used from boiler safety and superheater safctics and tcn inch escape piping from remaining steam systems. Low points in escape piping drain to the bilge.

Boiler Safety Valve Settings	Lift	Reset
#1 Drum, 2-1/2" Valve	750 psi	728 to 712 psi
#2 Drum, 2-1/2" Valve	745 psi	723 to 708 psi
Superheater Safety Valve1-1/2"	645 psi	626 to 612 psi (3 to 5%)

Initial Superheater Temperature Control

Protection against excessive superheater temperature while lighting off and bringing the boiler up to pressure is provided by means of a 1" superheater vent line, which permits the circulation of steam through the superheater to atmosphere through a root and throttle valve connection. These valves provide two valve superheater protection in the vent line to insure against steam leakage to atmosphere once the boiler is up to operating pressure.

Automatic Superheater Temperature Control

The superheater temperature control equipment has been removed from the Patriot State boilers, but is similar to equipment found on many ships, and is described here for reference only. The superheater outlet temperature was maintained by diverting steam from the inlet of the fourth superheater pass, through a superheating coil located in the boiler steam drum and then back to the inlet of the fifth superheater pass. The control equipment was capable of maintaining superheater outlet temperature within 5° F. of the set point during a superheater outlet temperature change of one °ee in three seconds. The control was normally set to maintain 905° F. at the superheater outlet. The controls consisted of a pneumatic temperature transmitter which sensed the superheater outlet temperature and transmited an air pressure proportional to this temperature to a pneumatic controller. The controller established an air pressure proportional to the difference between the superheater outlet temperature and a set temperature, and transmitted this pressure via a valve positioner to a disphragm operated valve to control the flow of steam through the control desuperheater. The valve positioner accurately and rapidly positioned the valve in response to the controller. A remote control station was provided to establish a set point for maintenance of the superheater outlet temperature.

A stop valve was provided at the outlet and inlet of the 4" superheater temperature control line and a 1/2" circulating by-pass around the diaphragm operated control valve. A control desuperheater controlled and maintained the superheater outlet temperature within a preset range. This was necessary when the boilers were operating near their maximum firing rate. The outlet temperature of an uncontrolled superheater would tend to rise according to the firing rate.

Since this ship, in its present condition, no longer has a control desuperheater we are left with an uncontrolled superheater. The superheat temperature must be watched closely and may be controlled to a certain extent by using three factors: excess air, feedwater temperature, and which burners we use. Too much excess air used for combustion will raise the superheat temperature, and therefore must be closely watched. Higher feedwater temperature entering the boiler will lower the superheater outlet temperature, so the third and fourth stage heaters should be utilized to their fullest potential. When operating the boilers with only two burners, it is undesireable to have the two inside burners firing. These burners, being closer to the superheater will transfer more heat to the superheater bank increasing the temperature.

Soot Blowers

Each boiler has the following steam soot blowers:

- 3 for the generating bank. (air driven)
- 2 IK retractable for the superheater. (air driven)
- 1 for hot gas side of the air preheater. (motor driven)
- 1 for the cold side of the air preheater. (motor driven)

Soot blowers enable the operating personnel to keep the boiler tubes free of soot deposits. The soot blowers are installed adjacent to and between the tubes. They are used while the boiler is steaming and act on the fire side of the tubes. Soot blowers are pipes with nozzles along their length. Soot blowers can either be permanent or retractable, and they can rotate up to a full 360° during operation. Steam for operation is supplied by the 600 psi auxiliary steam line. When secured, the piping in the nonretractable soot blowers is kept cool with the use of a steady flow of air passing through them. The air is supplied by the positive pressure in the boiler double casing. This air flow prevents deterioration by cooling the soot blower and also cleans, as well.

The nozzles on the soot blower element are so arranged that they do not spray the high velocity steam directly on the tubing which would cause erosion. The steam is cut in and out to the element by the soot blower head which has a valve operated by a cam and a rocker arm. The cam and element can be rotated by a hand operated endless chain or an air motor. Soot blowers are blown in a predetermined order, to follow the gas flow. The speed of the forced draft blowers are increased, when blowing tubes, to help produce a force of air to carry the soot up the stack.

A raise in stack gas temperature would indicate poor heat transfer in the boiler itself. Soot must be removed from the tube surfaces, at least daily, with the soot blowers. If the rotating element of the soot blowers is non-functioning, do not use the blowing unit. The stationary nozzle will direct steam in only one direction and severe tube erosion could occur.

Boiler Blow Down

Patriot State has no surface blow line installed. Bottom blow valves are installed at the bottom of each water drum and water heaters. The valves used in bottom and surface blow down are used to reduce total solids or grease, dirt or to just drain off excess water when shut down.

When blowing down a boiler, you must never allow the boiler level to go out of sight in the gauge glass. A boiler should never be given a bottom blow while steaming unless it is an emergency. This is because circulation could be interrupted. The most effective procedure for blow down of the boiler is to secure the boiler for an hour and then give a short blow down. This should be repeated every half hour until sufficient sludge is removed. A rest period between the multiple short blows will allow the sludge to settle each time and the blow down will be far more effective.

F.O. Burners

The burners onboard the Patriot State are manufactured by Babcock and Wilcox Company. They are steam atomizing type burners. There are four burners per boiler. The burner is made up of two principle components; the fixed air register and the removable atomizer assembly. The atomizer sprays the oil into the furnace in a very fine mist by the use of steam which is the sole atomizing force. Variations in capacity are obtained by increasing or decreasing the oil pressure being supplied to the atomizer and thence to the oil flow paths which intersect with the atomizing steam paths within the sprayer plate. Accordingly, steam must be used the entire time the steam atomizing plate is in service.

The oil under header supply pressure and temperature enters the shut-off coupling at the valve body oil inlet connection on the left-hand side of the coupling. The oil is then directed to the annular space between the outer barrel and the internal tube. The oil and steam, therefore, follow separate paths until they reach the sprayer plate. The intersection of the oil and steam within their respective exit paths results in atomization of the fuel being wholly completed within the exit passages. Breaking up the oil spray into a number of discharge jets provides greater surface exposure of the fuel particles to the surrounding air for combustion and results in good combustion over the full capacity range of the burner provided proper oil - air ratio is maintained.

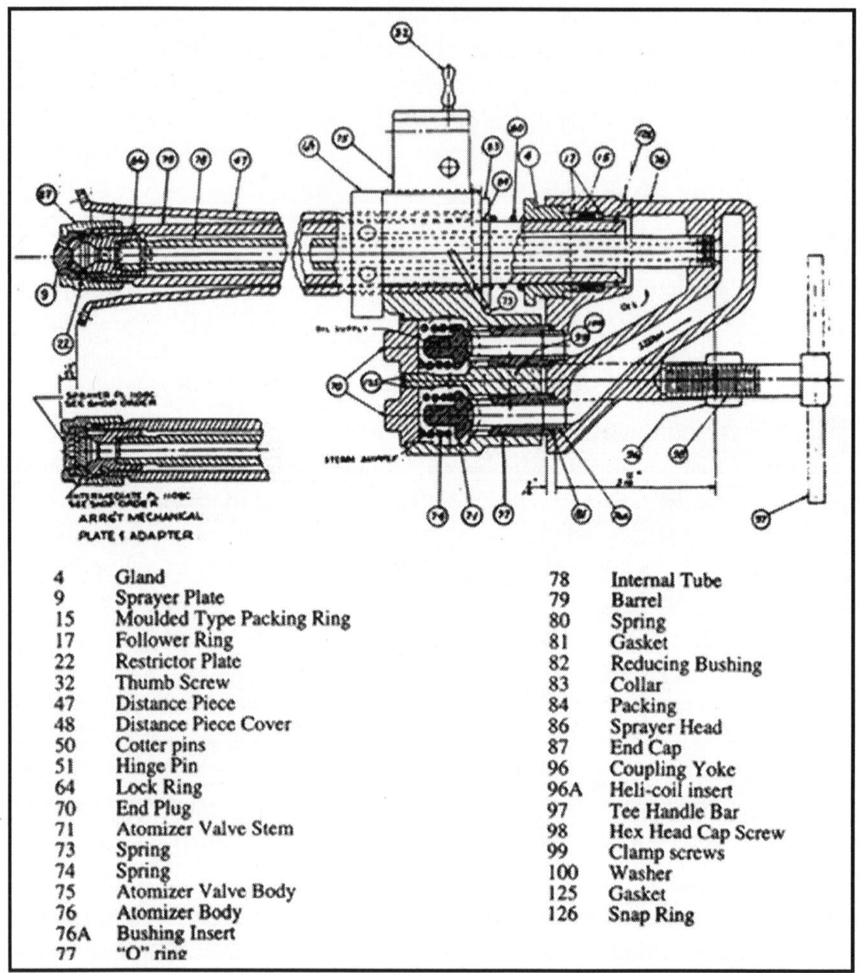

Fig. Y-Jet Atomizer.

Boiler Accessaries

Each boiler is fitted with two gage glasses, high and low water level indicators, a draft gage, exhaust temperature pyrometers, and a smoke periscope to provide necessary operating information to the fireman. The upper gage glass is of the standard type, while the lower glass is a bicolor type. A bicolor glass operates upon the optical principle of refraction. A beam of light passes through the gage and then through a red or green filter. The different refraction properties of light when passing obliquely through water or steam determines which filter is illuminated. In a bicolor gage glass, water appears green and steam appears red.

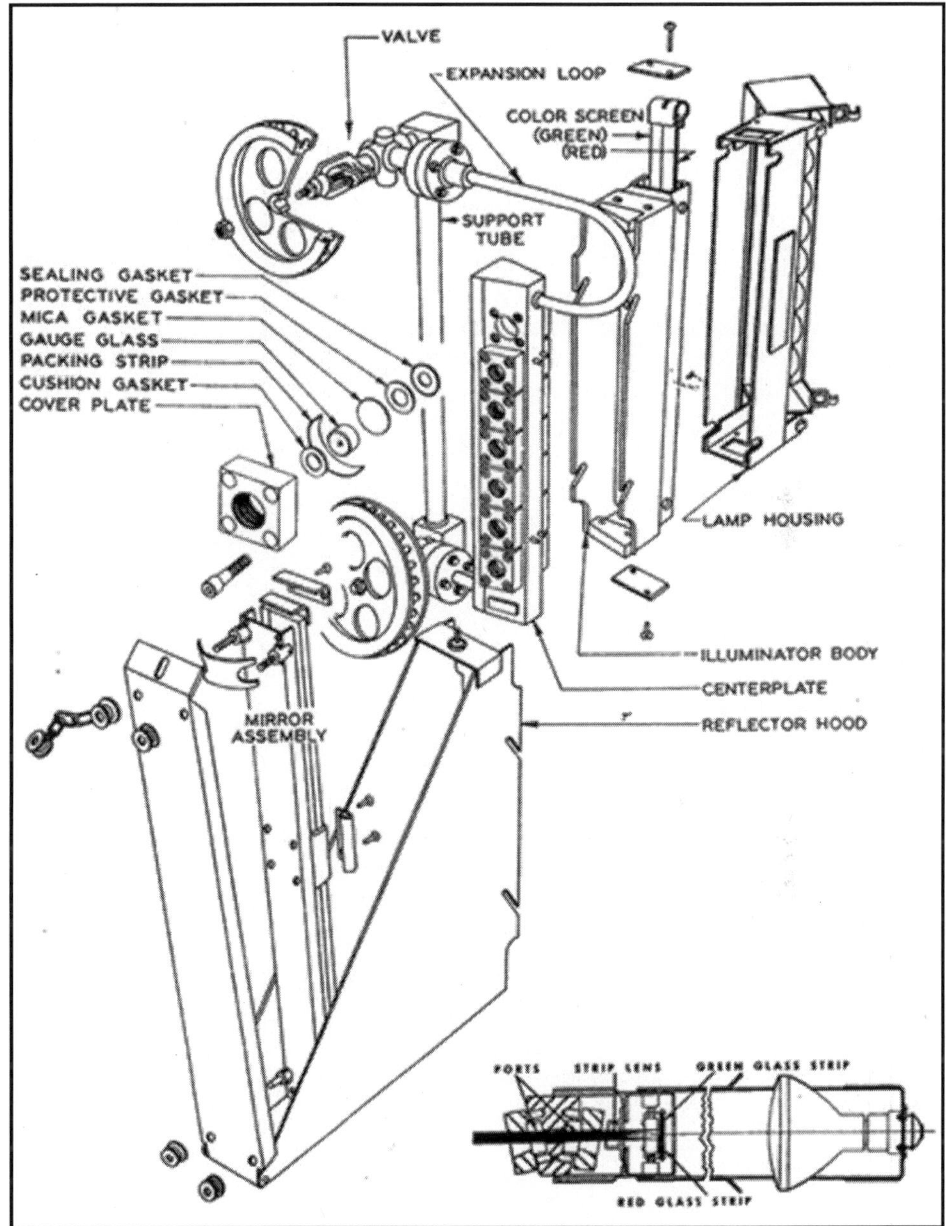

Fig. Bi-color Gauge Glass.

The smoke periscope is a simple arrangement of mirrors and a light bulb which shines across the uptakes, giving the operator an indication of the opacity of the combustion gasses.

It is difficult to destinguish between white and black smoke with the periscope.

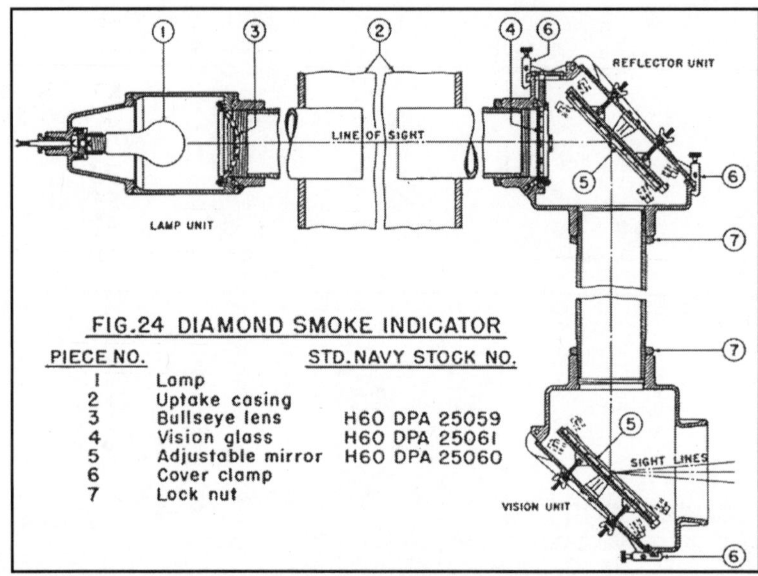

Periscope

SCOTCH MARINE BOILER

Fig. German example. Note the steam dome, a typically German feature, and also the corrugated furnaces.

A "Scotch" marine boiler (or simply Scotch boiler) is a design of steam boiler best known for its use on ships. The general layout is that of a squat horizontal cylinder. One or more large cylindrical furnaces are in the lower part of the boiler shell. Above this is a large number of small-diameter fire-tubes. Gases and smoke from the furnace pass to the back of the boiler, then return through the small tubes and up and out of the chimney. The ends of these multiple tubes are capped by a smokebox, outside the boiler shell.

The Scotch boiler is a fire-tube boiler, in that hot flue gases pass through tubes set within a tank of water. As such, it is a descendant of the earlier Lancashire boiler and like the Lancashire it uses multiple separate furnaces to give greater heating area for a given furnace capacity. It differs from the Lancashire in two aspects: a large number of small diameter tubes (typically 3 or 4 inches (76 or 102 mm) diameter each) are used to increase the ratio of heating area to cross-section. Secondly the overall length of the boiler is halved by folding the gas path back on itself.

COMBUSTION CHAMBER

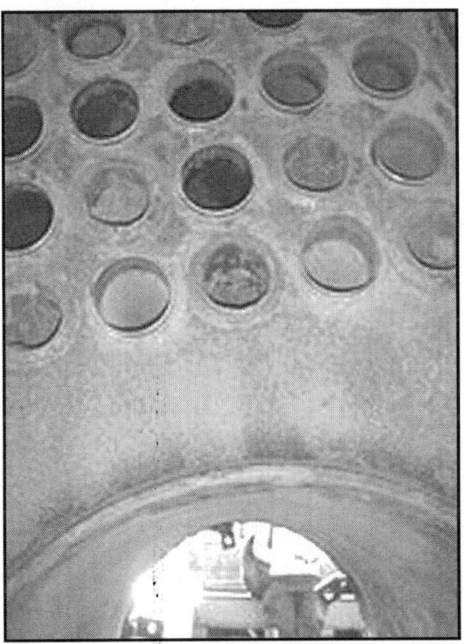

Fig. Inside the combustion chamber, looking up at the tubeplate.

Fig. Rear face of the boiler of steam tugMayflower, showing the stays supporting the combustion chambers.

The far end of the furnace is an enclosed box called the combustion chamber which extends upwards to link up with the firetubes. The front wall of the combustion chamber is supported against steam pressure by the tubes themselves. The rear face is stayed by rod stays through the rear shell of the boiler. Above the combustion chamber and tubes is an open steam collecting space. Larger long rod stays run the length of the boiler through this space, supporting the ends of the boiler shell.

With multiple furnaces, there is a separate combustion chamber for each furnace. A few small boilers did connect them into one chamber, but this design is weaker. A more serious problem is the risk of reversing the draught, where exhaust from one furnace could blow back and out of the adjacent one, injuring the stokers working in front of it.

ORIGINS

The first recorded boiler of comparable form was used in a railway locomotive, Hackworth's 'Wilberforce' class of 1830. This had a long cylindrical boiler shell similar to his earlier return-flued 'Royal George', but with the return flue replaced by a number of small firetubes, as had been demonstrated so effectively by Stephenson with his 'Rocket' a year earlier. The novel feature of an entirely internal combustion chamber was used. Unlike the later Scotch boiler though, this was self-supported by its own stays, rather than using stays through the walls of the boiler shell. This allowed the entire assembly of outer tubeplate, furnace tube, combustion chamber and firetubes to all be removed from the boiler shell as one unit, simplifying manufacture and maintenance. Although a valuable feature, this became impractical for larger diameter chambers that would require the support of the shell.

VARIANTS

Number of Furnaces

Typical practice for ships was to have two furnaces in each boiler. Smaller boilers might only have one, larger boilers commonly had three. The limitation in boiler size was the amount of work each stoker could do, firing one furnace per man. Larger ships (meaning anything above the smallest) would have many boilers. As with the Lancashire boiler, the furnace was often corrugated for strength. Various makers had their own particular ways of making these corrugations, leading to their classification for maintenance purposes under the broad titles of, Leeds, Morrison, Fox, Purves or Brown.

Wet Back and Dry Back

The typical design is the "wet back", where the rear face of the combustion chamber is water-jacketed as a heating surface. The "dry back" variation has the rear of the combustion chamber as an open box, backed or surrounded only by a sheetmetal jacket.This simplifies construction, but also loses much efficiency. It is only used for small boilers where capital cost outweighs fuel costs. Although the Scotch boiler is nowadays rarely the primary steam generator on a ship, small dry-back designs such as the Minipac are still encountered, for supporting secondary demands whilst alongside in port with the main boilers cold.

One interesting variant of the dry-back design has been a patent for burning ash-prone fuels. The rear of the combustion chamber is used as an access point for an ash separator, removing the ash before the small-diameter tubes.

Double-Ended

The double-ended design places two boilers back-to-back, removing the rear wall of the boiler shell. The combustion chambers and firetubes remain separate.

This design saves some structural weight, but it also makes the boiler longer and more difficult to install into a ship. For this reason they were not commonly used, although back-to-back arrangements of multiple single-ended boilers were common.

Inglis

The "Inglis" modification adds an extra combustion chamber where an additional single large flue returns from the rear to the front of the boiler. Flow through the multiple tubes is thus from front to back, and so the exhaust is at the rear.

Multiple furnaces would share a single combustion chamber. The major advantage of the Inglis is the extra heating area it adds, for a comparable shell volume, of perhaps 20%. Surprisingly this is not from the additional combustion chamber, but from lengthening the narrow firetubes. These can now run the full length of the boiler shell, rather than just the rather shorter distance from the inner combustion chamber to the front tubeplate. Despite this advantage, it is rarely used.

USE IN SHIPS

The Scotch marine boiler achieved near-universal use throughout the heyday of steam propulsion, particularly for the most highly developed piston engines such as the triple-expansion compounds. It lasted from the end of the low-pressure haystack boilers in the mid-19th century through to the early 20th century and the advent of steam turbines with high-pressure water-tube boilers such as theYarrow.

Large or fast ships could require a great many boilers. The Titanic had 29 boilers: 24 double-ended and 5 smaller single-ended. The larger boilers were

15 feet 9 inches (4.80 m) diameter and 20 feet (6.1 m) long, the smaller were 11 feet 9 inches (3.58 m) in length. All had three corrugated Morrison furnaces of 3 feet 9 inches (1.14 m) diameter, 159 furnaces in total, and a working pressure of 215 pounds per square inch (1,480 kPa).

Shipboard Working Examples

Numerous Scotch boilers are in use on ships as of 2011, and new boilers can be built to replace life-expired ones. Examples of preserved steam boats employing Scotch boilers include:

- Steam tug Mayflower, Bristol Industrial Museum
- Mayflower's boiler was recently removed for restoration and could easily be seen in close-up.
- Baltimore, of 1906, at the Baltimore Museum of Industry, is the oldest working steam tug in the US.
- Steamship Shieldhall based in Southampton, UK, is fully operational and has two oil-fired Scotch boilers.
- Steam-powered icebreaker Stettin, operating on two coal-fired boilers
- Steam-powered drifter Lydia Eva, based in Lowestoft or Great Yarmouth, Fully operational coal-fired Scotch boiler.
- The steam tug Kerne built 1913. Operated out of Liverpool by the Steam Tug Kerne Preservation Society. One coal-fired wet back Scotch marine boiler with two furnaces. Max working pressure 180 pounds per square inch (1,200 kPa).
- The steam yacht Louise built 1902. Operating on Lake Geneva, Wisconsin by Gage Marine Corporation. One diesel-fired Scotch marine boiler powering a double-expansion steam engine, in excursion passenger service.
- The steam ship Trafik built 1892 in Stockholm Sweden by Bergsunds Mekaniska Verkstad as a passenger and freight ship on the Swedisk lake Vättern and with homeport Hjo. The machinery and its boiler are the original ones made for the ship in Stockholm.

PROCEDURE FOR STARTING AND STOPPING A BOILER

Twitter Google+ LinkedIn Pinterest

A boiler is one of those machineries that gets the ship going. A boiler is something, which though not required continuously in operating a ship, cannot be done away with. Moreover, it's a dangerous equipment which generates steam at extremely high pressure, and it is for this reason that proper care should be taken while operating it.

In this article we have brought to you a step-by-step procedure for starting and stopping a boiler on a ship. With this procedure you can never go wrong, as far as boilers are concerned. Starting and stopping a boiler was never so easy.

Starting a Boiler

It is to note that the following steps may not apply to all types of boilers and each boiler requires some additional steps to be followed as per its system design.

However, the basic steps remain the same:

1. Ensure that the vent valve on the boiler is open and check there is no pressure in the boiler.
2. Check that the steam stop valve is closed.
3. Check that all the valves for fuel are open, and let the fuel circulate through the system until it comes to the temperature required by the manufacturer recommendation.
4. Check and open the feed water valves to the boiler and fill the water inside the boiler drum to just above the low water level. This is done because it is not possible to start the boiler below the low water level due to safety feature which prevent boiler from starting. Also, the level is not filled much because if filled too much, the water inside the boiler might expand and over pressurize the boiler.
5. Start the boiler in automatic mode. The burner fan will start the purging cycle which will remove any gases present in the furnace by forcing it out through the funnel.
6. After the pre-set purge time the pilot burner will ignite. The pilot burner consists of two electrodes, through which a large current is passed, via the transformer, producing the spark between the electrodes. The pilot burner is supplied with diesel oil and when the oil passes over, the former ignites.
7. The main burner which is supplied by heavy oil catches fire with the help of pilot burner.
8. Check the combustion chamber from the sight glass to ensure the burner has lit and the flame is satisfactory.
9. Keep a close eye on the water level as the pressure increases and

open the feed water when the level of water inside the gauge glass is stable.
10. Close the vent valve after the steam starts coming outside.
11. Open the steam stop valve.
12. Once the working steam pressure is reached, blow down the gauge glass and float chambers to check for the alarms.

Stopping a Boiler

1. If the boiler is needed to be stopped for longer duration for maintenance or opened up for survey, change the fuel to distillate fuel.
2. If separate heating arrangement for heavy oil is present then there's is no need to change over to distillate fuel and the oil is kept on circulation mode.
3. Stop the boiler automatic cycle.
4. Close the steam stop valves.
5. Close the boiler feed water valves.
6. When the boiler pressure is just reduced to over atmospheric pressure the vent valve is kept open to prevent vacuum formation inside the boiler.

BASIC CONSTRUCTION

This design allows the use of lower quality feed. It is easy to clean and easy to maintian and replace tubes. Disadvantages are the large number of handhole doors and the extensive brickwork. The drum is all welded and the casing bolted

FRONT FIRED STUDDED WALL REFRACTORY COVERED BOILER

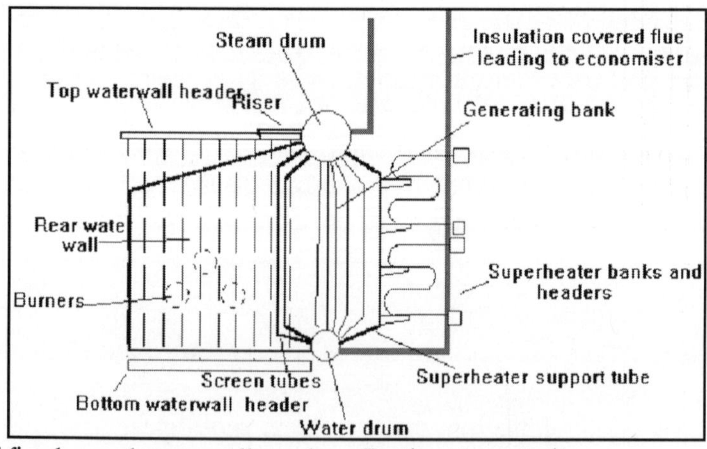

Roof fired membrane wall modern Radiant heat boiler

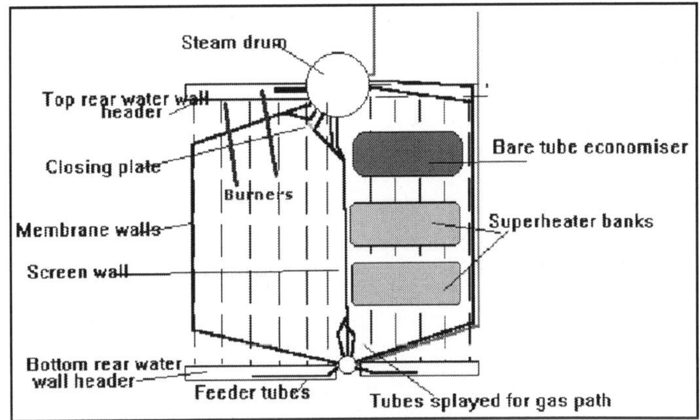
Steam drum
Top rear water wall header
Closing plate
Burners
Membrane walls
Screen wall
Bottom rear water wall header
Feeder tubes
Tubes splayed for gas path
Bare tube economiser
Superheater banks

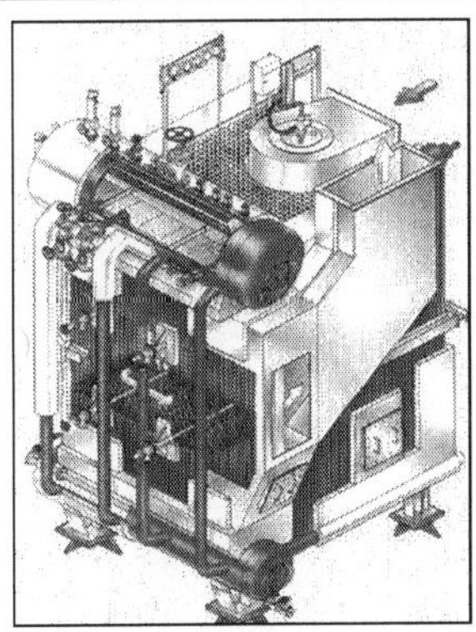

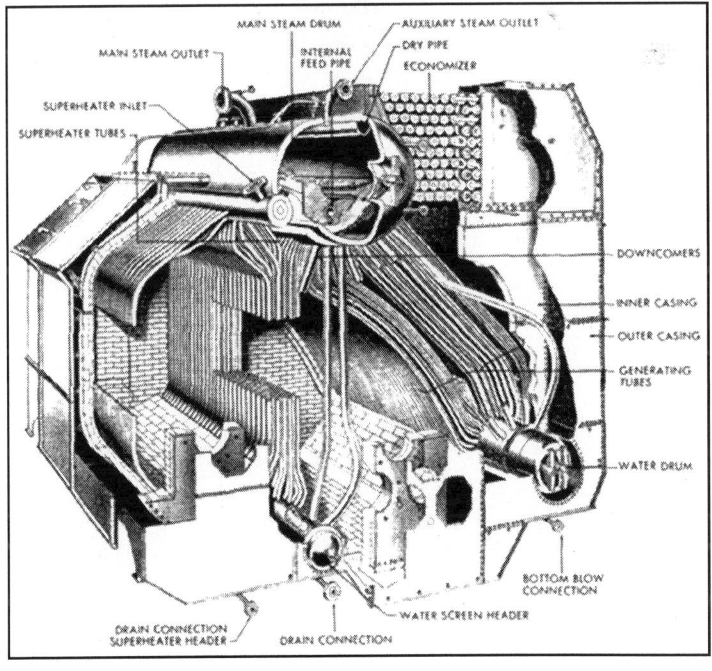
MAIN STEAM DRUM
AUXILIARY STEAM OUTLET
DRY PIPE
MAIN STEAM OUTLET
INTERNAL FEED PIPE
ECONOMIZER
SUPERHEATER INLET
SUPERHEATER TUBES
DOWNCOMERS
INNER CASING
OUTER CASING
GENERATING TUBES
WATER DRUM
BOTTOM BLOW CONNECTION
WATER SCREEN HEADER
DRAIN CONNECTION SUPERHEATER HEADER
DRAIN CONNECTION

COMPONENTS

Steam Drum

In the early designs the drums were riveted or solid forged from a single ingot, but for modern boilers the drum is generally fabricated from steel plate of differing thicknesses and welded. The materials used are governed by classification society rules. Test pieces must be provided. The cylindrical drum is normally constructed from four plates. Two dished End plates, a thick wall tube plate (thicker to accommodate the holes drilled in it without increased stress) and completed with a thinner wrapper plate.

Construction takes the form of rigidly clamping the descaled, bent wrapper and tube plates together. In addition test pieces cut from the original material are attached to the construction in such away that the longitudinal weld extends either sided of the join. These pieces are later removed and shaped test shapes cut out from specified areas including across the weld. The longitudinal weld is critical (taking twice the circumferential stress) and is normally carried out by specialised automatic machinery using submerged arc techniques. The dished end pieces are accurately aligned and welded. On completion the construction is cleaned and non-destructive testing- such as x-ray photography, carried out. Final machining is carried out and any stub pieces and doublers attached. The now complete drum is heat treated at 600 to 650'C.

The final process is hydraulic testing to classification requirements. Natural circulation within a boiler is due to the differing specific gravities of the water at the differing temperatures, the steam drum provides a reservoir of cool water to give the gravitational head necessary for natural circulation. Cool water entering the steam drum via the feed lines provides the motive effect for the circulation distributing it to the downcomers. Also the space within the drum provides for the separation of the steam and water emulsions formed in the water walls and the generating tubes. Water droplets entrained with the separated steam are removed by separating components fitted in the drum as well as the perforated baffle plates fitted at the water line. The space above the water line provides for a reserve steam space needed to maintain plant stability during manoeuvring conditions. Also fitted are the chemical injection distributing pipe and the scuming plate.

The smaller the drum is made, the less thickness of material that is required. However, the limitation to how small is that sufficient space must be allowed for the separation of water from the steam before passing out to the superheater space otherwise dryers must be used. Also, due to the smaller reserve of water, larger fluctuations in water level occur during manoeuvring.

WATER DRUM

Distributes feed water from the downcomers to the headers and generating tubes. Provides a space for accumulating precipitates and allows them to be blown down. Water drum size is limited to that required to receive the generating tubes, for modern radiant heat boilers with only a single bank of screen tubes and no generating tubes between the drums, the water drum has been replaced by a header and the downcomers fed straight to the waterwall headers. With system blow down is done at the steam drum. Too small a water drum can cause problems of maintaining ideal water level and little steam reserve

HEADERS

These have a similar purpose to the water drum but are smaller in size. Due to their reduced size they may have a square cross section without resorting to exceptional thickness. .

GENERATING TUBES

Consists of a large number of small diameter tubes in the gas flow, more commonly found in boilers of an older design For roof fired boilers the generating bank may consist of one or two rows of close pitched tubes. For a modern radiant heat boiler the generating bank has been omitted to allow the replacement of the water drum by a distribution header, a bare tube economiser is fitted generating 5% of the steam capacity. The generation bank is normally heated by convection rather than radiant heat.

For a set water circulation the tube diameter is limited to a minimum as the ratio of steam to water can increase to a point where the possibility of overheating could occur due to the lower heat capacity of the steam. The number of tubes is limited to prevent undercooling of the gas flow leading to dew point corrosion

SCREEN TUBES

These are larger bore tubes receiving the radiant heat of the flame and the convective heat of the hot gasses. The large diameter keeps the steam/ water ratio down hence preventing overheating. There main duty is to protect the superheater from the direct radiant heat. On a modern marine radiant heat boiler the screen wall is formed out of a membrane wall

WATERWALL TUBES

Contains the heat of the heat of the furnace so reducing the refractory and insulation requirements.

Comes in four designs

- Water cooled with refractory covered studded tubes
- Close pitched exposed tubes
- Monowall
- Membrane Wall

Mono and membrane walls greatly improve on the older designs in being gas tight

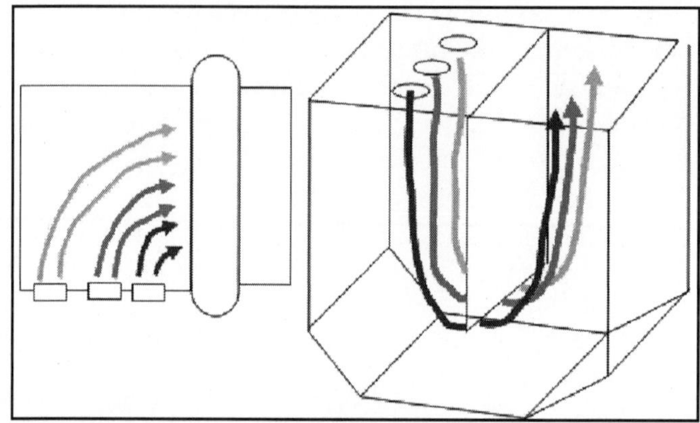

Advantages of roof firing over side firing

- Increased efficiency due to the longer length allowed for the flame giving more time for complete combustion. This also allows more heat to be released as radiant rather than convective cutting down the required number of screen wall generating tubes
- The longer period allowed for complete combustion means that less excess air is required, this has the knock on effect of lowering the Dew Point of the flue gasses.
- Equal length flames
- Better gas flow
- For roof fired the effect of each flame is the same, foir side firing it differs. To keep within the design limitations the boiler must be operated to the highest effect flame with the other two operating at reduced effect

Ligament (ligature) Cracking Mechanics

Generally associated with failure of refractory plug located beneath steam drum. The ligature is the space between the tube plate holes. Classification rules typically allow isolated ligature cracks to be gouged and re-welded. For continuous cracks repairs are not normally allowed and a new drum/tube plate may be required

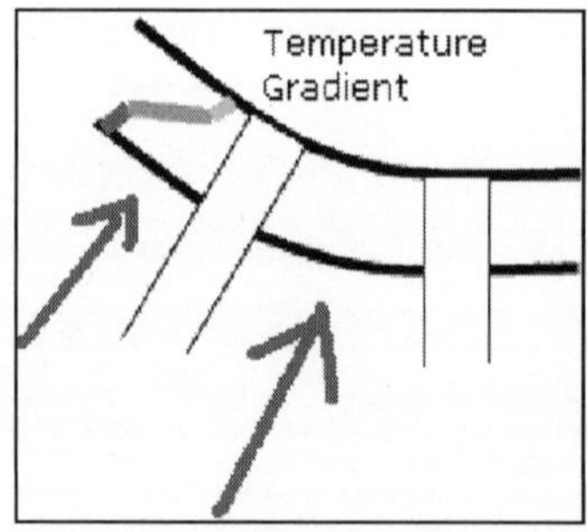

Hot gasses acting on the thick section tube plate set up a temperature gradient leading to creep, plastic flow to relief thermal stress and high tensile stress on the surface at cool down. In addition grain growth leads to the metal becoming brittle

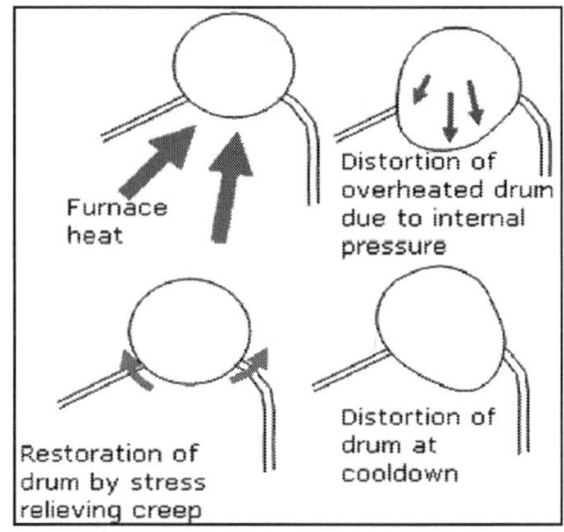

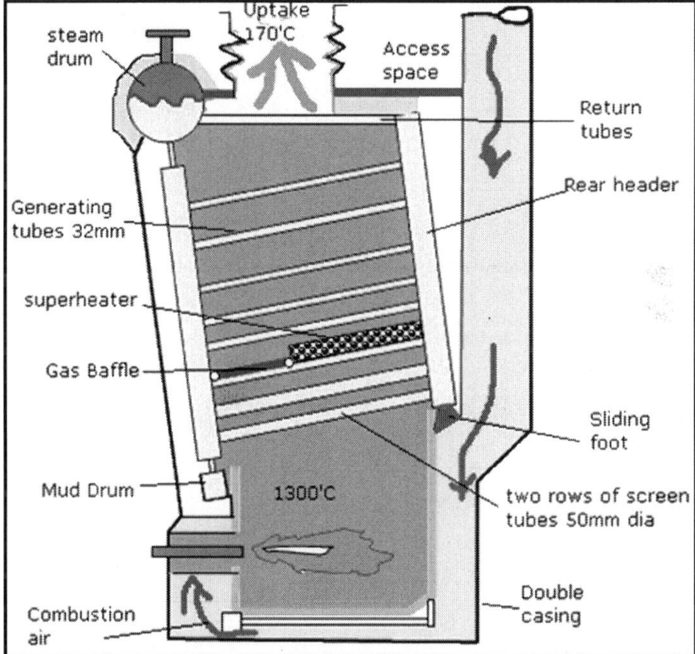

A more severe form may lead to distortion of the entire drum in two possible directions. The thick section tube plate is exposed to the heat of the furnace and is subject to overheating. Thermal distortion takes place leading to stressing.

This stressing is relieved by creep . When the drum cools a set distortion is in place The distortion may occur in three ways, in a radial or axial direction as shown.

ADVANTAGES OF WATER TUBE BOILERS OVER SMOKE TUBE (TANK)

Advantages Over Tank

- Savings in weight of about 3:1 for a comparable heating surface area
- Possibility of using higher temperatures and pressures without unduly increasing wall thickness increases plant efficiency.
- More efficient combustion space allowed
- Greater flexibility of the structure and rapid circulation prevents the problems of thermal stressing in the tank boilers which leads to grooving. In water tube boilers roof and floor tubes are sloped at 15' to ensure circulation
- thinner tube materials allow rapid steam raising and faster heat transfer rates
- Saving in space for same steaming rate
- Wider safety margins- limited tube diameters and protected drum surfaces mean failure in tubes releases a flow of steam dependent on tube diameter
- Thin tubes are easier to bend, expand and bell mouth

Disadvantages

- Lower reserve of water means a more efficient water level control is required
- High quality feed required
- little allowance to corrosion

DOWNCOMERS

These are large diameter unheated i.e. external to the furnace, their purpose is to feed water from the steam drum to the water drum and bottom headers.

RISER/RETURN TUBES

These return steam from the top water wall headers to the steam drum.

SUPERHEATER TUBES

These are small diameter tubes in the gas flow after the screen tubes. Due to the low specific heat capacity of the saturated steam they require protection from overheating in low steam flow conditions, say when flashing.

SUPERHEATER SUPPORT TUBES

These are large diameter tubes designed to support part of the weight of the superheater bank of tubes.

MATERIAL REQUIREMENTS

Tube temperatures for the water cooled sections is considered to be saturation temperature plus 15oC. Solid drawn mild steel is generally used. Tube temperatures for convection superheater sections is considered to be final superheat temperatures plus 30oC. For Radiant heat a higher temperature is considered. For Superheater tubes operating above 455oC a Chrome Molybdenum alloyed steel is required.

ADVANTAGES OF MEMBRANE/MONOWALLS

These were originally introduced in land power stations after experience had been gained in making the lower parts of the furnace sufficiently tight to hold liquid ash. This was achieved by welding steel strips between the floor tubes. Further development resulted in completely gas tight furnace wall panels being constructed by welding together either finned tubes or normal plane tubes with steel strips in between and welded. In both methods he longitudinal welds are done by automatic processes and panels of the required size are built up in the factory ready for installation into the boiler in one piece.

- Entire walls may be prefabricated
- Maintenance costs, particularly of insulation are lower
- Lower quality fuels may be used due to the much reduced amount of insulation reducing problems of slagging
- Simplified water washing procedures
- Due to gas tight seal there is no corrosion of outer casing.

A disadvantage would be that tube replacement following failure is more difficult. Also, the possibility of entire walls parting from the drum can occur during a furnace explosion.

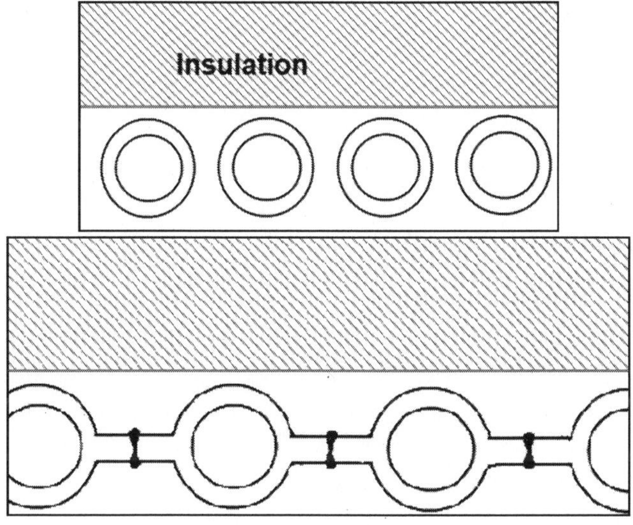

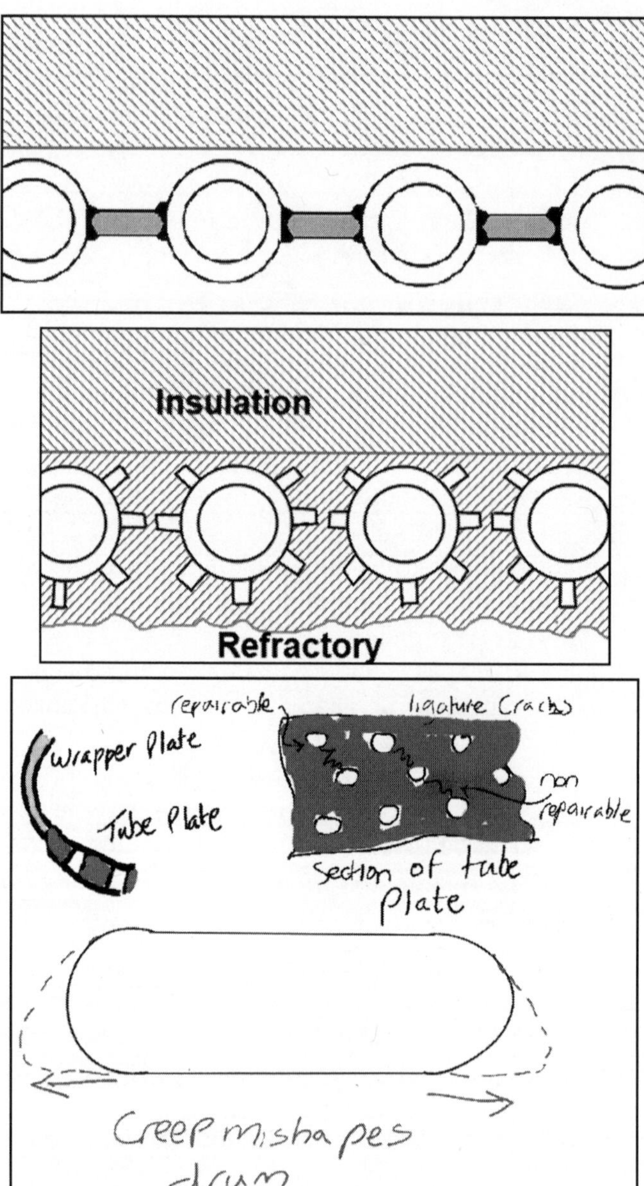

SMOKE TUBE BOILER

These were the most common form of boiler design before the introduction of water tube designs. See Comparisons of water tube and Smoke tube boilers. This style of boiler still see active service were low quantities of low quality steam are required, such as for cargo and fuel tank heating when in port. This style of boiler is relatively cheap, supplied as a packaged unit and requires less stringent feed water conditioning and level control.

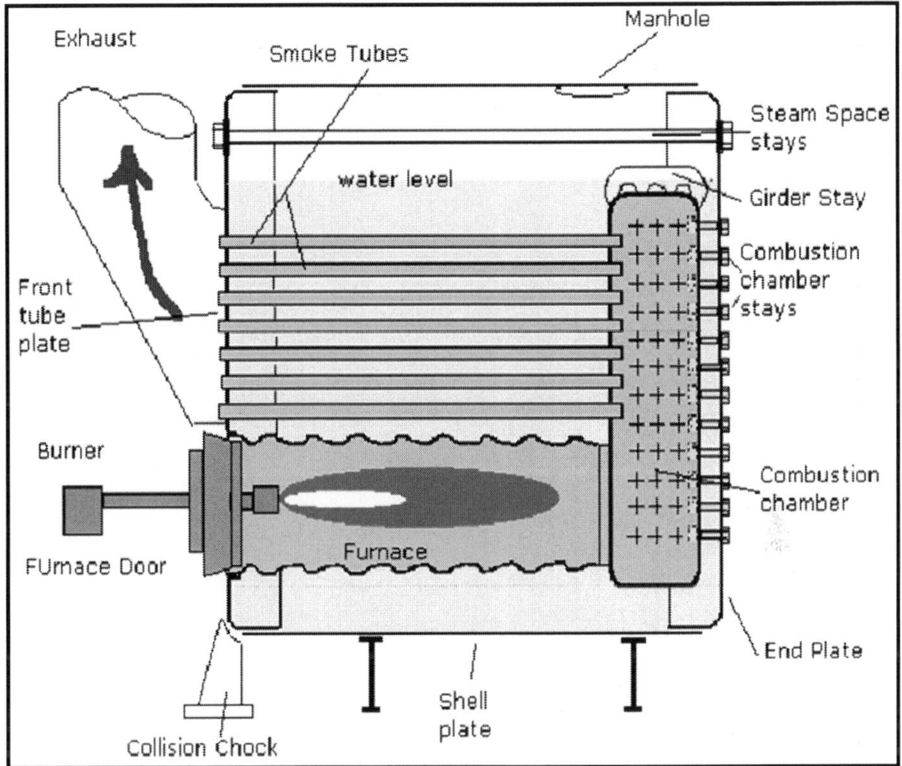

DESIGN

Consists of a shell wrapper plate to which is welded (or for later designs riveted, end plates.. Pressure is naturally container in the shell plate due to is cylindrical design.

The flat end plates, however, must be 'stayed' to prevent buckling and distortion. The combustion chamber is of similar section and is also 'stayed'.

The boiler shown above is a single furnace, two pass design. Larger boilers may have multiple furnaces and have multiple passes by replacing the exhaust stack with a return chamber and fitting another bank of tubes. The smoke tubes may be plain or threaded to act as stays. There are one stay tube for every three plain tubes approx.

These were the most common form of boiler design before the introduction of water tube designs.

See Comparisons of water tube and Smoke tube boilers. This style of boiler still see active service were low quantities of low quality steam are required, such as for cargo and fuel tank heating when in port. This style of boiler is relatively cheap, supplied as a packaged unit and requires less stringent feed water conditioning and level control.

DESIGN

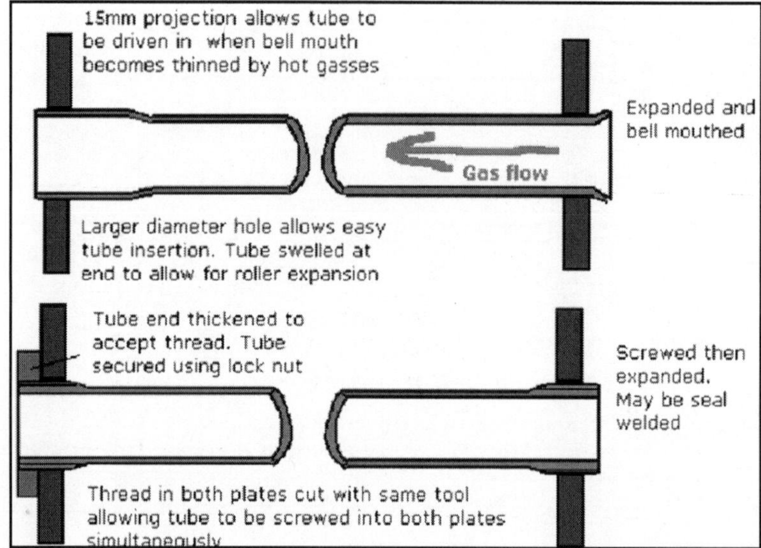

Consists of a shell wrapper plate to which is welded (or for later designs riveted, end plates.. Pressure is naturally container in the shell plate due to is cylindrical design. The flat end plates, however, must be 'stayed' to prevent buckling and distortion.

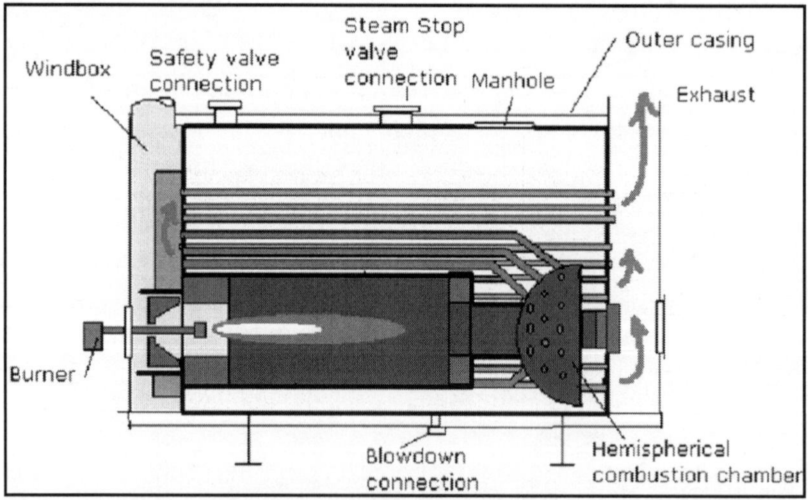

The combustion chamber is of similar section and is also 'stayed'. The boiler shown above is a single furnace, two pass design. Larger boilers may have multiple furnaces and have multiple passes by replacing the exhaust stack with a return chamber and fitting another bank of tubes. The smoke tubes may be plain or threaded to act as stays. There are one stay tube for every three plain tubes approx.

OPERATION

Although the maintenance of the water level is not so critical as with water tube designs, it should not be allowed to fall too much as overheating of furnace and combustion spaces leads to catastrophic failure due to component collapse. The content of the boiler is then expelled via the furnace door. Similarly, although water treatment is not so critical scale must not be allowed to build up which can lead to overheating of material

MARINE RADIANT REHEAT BOILERS

To increase plant efficiency reheat systems are used. In this the exhaust from the HP Turbine is led back to the furnace and reheated to superheat temperatures. This allows the steam to be expanded to lower pressures in the LP turbine with reduced need for blade taper twist and other efficiency degrading designs to cope with the steam wetness.

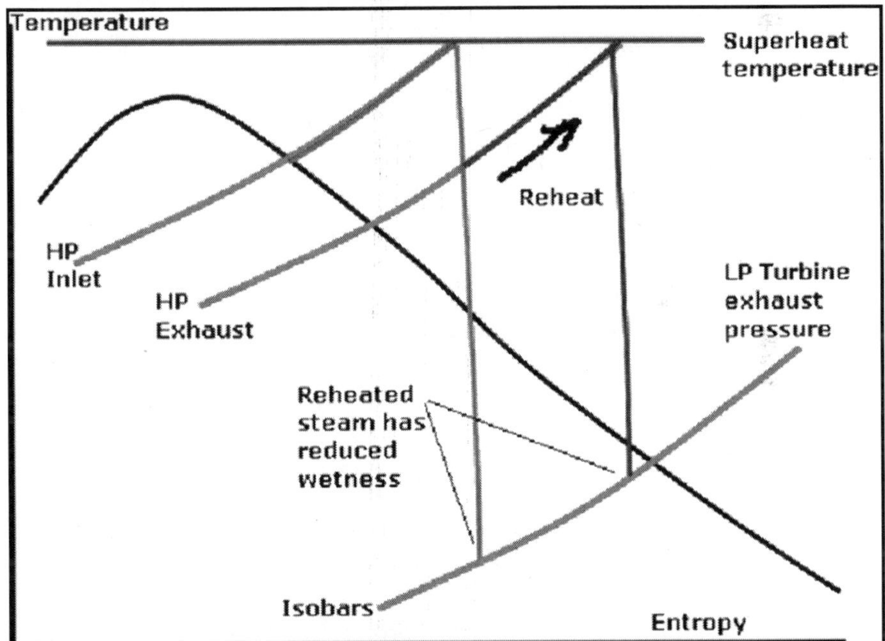

The boiler design is of a standard roof fired radiant furnace with a gas tight membrane water wall and single row of screen tubes. The convection space is divided into a superheater and reheater section and a section containing superheat and reheat temperature control by-pass economisers. Gas dampers allow cooling air from the windbox to pass over the reheater section during astern manoeuvres to prevent overheating and thermal shocking when the plant is moved ahead. An alternate design using water drum attemperation for superheat control rather than by-pass economisers splits the convective zone into two parts.

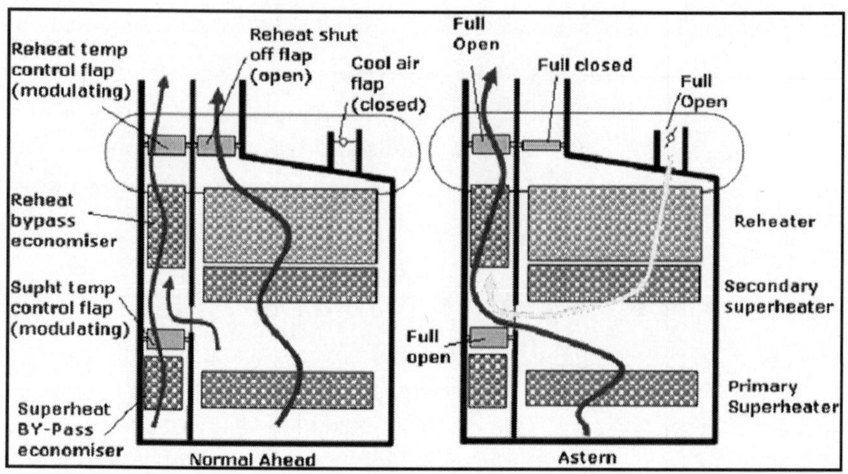

The superheaters in the reheat section ensure that gas is cooled sufficiently to prevent overheating in the reheat in the event of gas dampler leakage. Provision is made to allow windbox air into the reheat space in the event of damper failure.

Furnace Gas tight

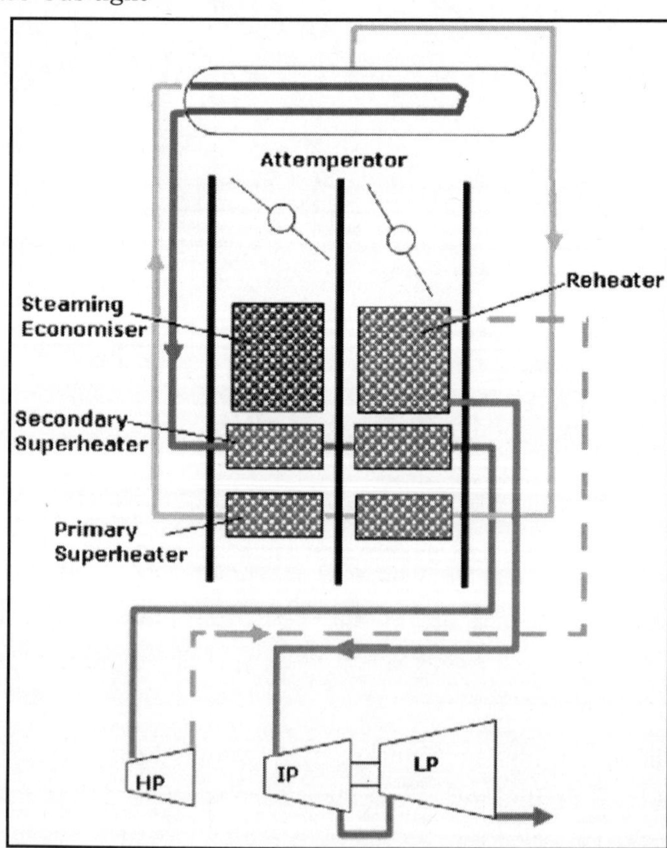

- Advantages of MRR boiler. These are common with any radiant boiler over the convective type design
- Increased plant efficiency
- Improved combustion
- Less excess air requirement. This has the additional advantage of reducing dew point corrosion in the uptake
- Refractory limited to burner quarls and exposed section of water drum and bottom header. This allows for poor quality fuels due to reduced slagging.

Dual Pressure (Double Evaporation) Boiler

The main reason for the adoption of this design of boilers is to allow use of modern high efficiency watertube boilers without fear of damage through contamination by cargo or fuel oils. The basic design consists of a D-Type boiler design upon which is mounted a Steam/Steam generator drum. The steam generated by the main boiler heats water in the Steam/Steam generator which produces steam requirements.

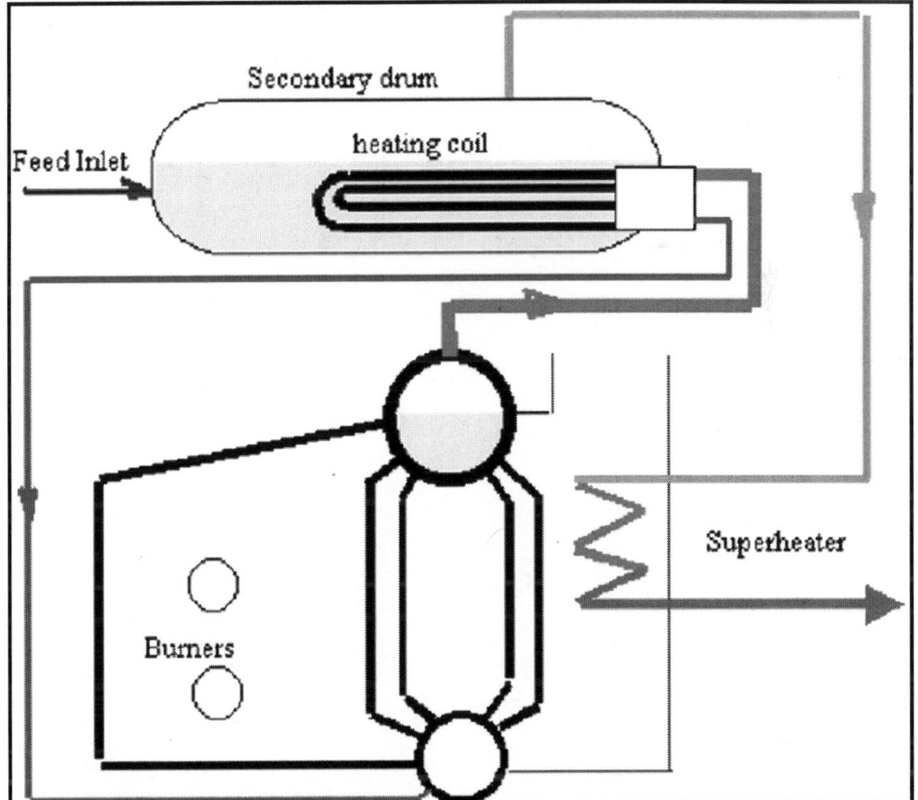

The primary drum is initially filled with high quality feed water and suitably dosed. Make up is limited to small amounts due to leakage therefore the feed pump may be of simple design. An example could be a steam or air driven reciprocating pump. The chemical treatment is simple with little requirement for addition or blowdown. The above design shows the fitting of a superheater. These are normally only fitted where the generated steam will be required to power turbine operated machinery most typically an alternator.

SECONDARY DRUM.

The U-tube heating elements are passed through the manhole dorr and expanded into headers welded into the dished end of the drum. The tubes are well supported.

A manhole may be fitted at the lower part of the shell allows access to the heating elements. The drum is generally mounted integral, supports are attached to the structure of the primary boiler. The secondary drum also acts as a steam receiver for the exhaust gas boiler.Typical pressures are 63 bar for the primary circuit and 23.5 for the secondary.

The author has not sailed with pressures anywhere near this with this design. Primary pressures of 35bar and secondary pressures nearer 15 bar have proven sufficient even to drive an alternator. Of note is that these designs are obviously more expensive than a normal single steam drum plant even taking into account the improved efficiency.

They are therefore generally associated with larger motor powered plant with large waste heat units capable of supplying all requirements including an alternator. However the author has sailed on this plant on a 20,000 tonne product tanker.

Where these boilers are installed in Motorships a “simmering coil” may be fitted.

This is located in the primary drum and is supplied from the exhaust economiser to keep both circuits warm thereby preventing any possible damage due to lay-up. Mountings are those typically found on any boiler with low level water alarms and low/low level shut off on both boilers. The accumulation of pressure test for the safety valves fitted to the secondary drum are calculated with the primary boiler firing at maximum rate generating maximum heating steam supply.

TYPICAL CIRCUIT INCORPORATING DUAL PRESSURE BOILER

Under port conditions the main boiler is fired to providing heating steam for the secondary drum. From this steam is supplied for tank heating or to a turbo-alternator via a superheater. When the vessel is underway the main boiler may stop firing.

A waste heat circulating pump passes water from the secondary drum via the waste heat unit back to the drum. The steam produced is again available for tank heating and powering a turbo-alternator. Cross over valves are fitted for Harbour and sea-duty conditions.

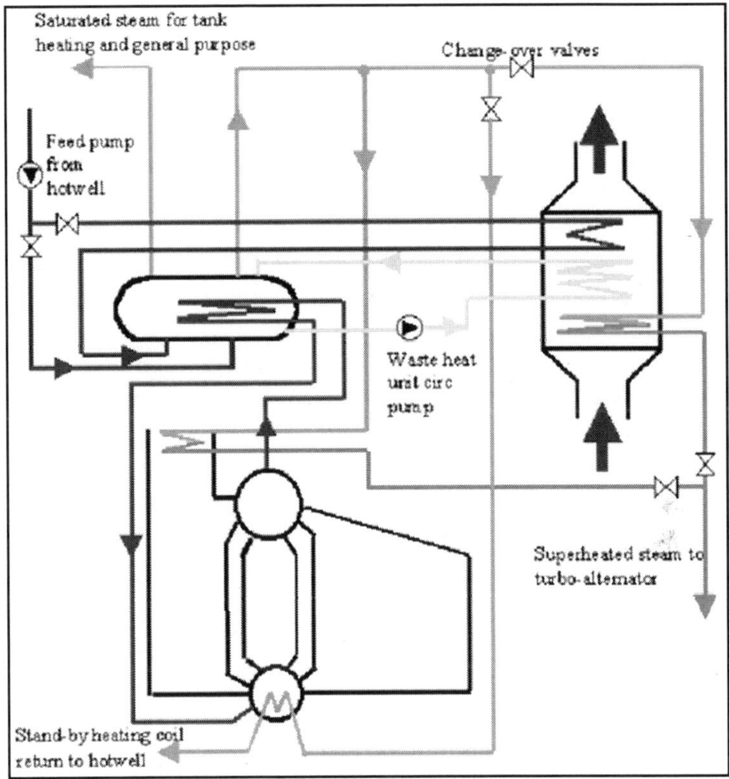

COMBUSTION CONTROL

The purpose of the combustion control system is to maintain constant steam pressure at the superheater outlet header for all loads by varying the rate of combustion in boiler furnaces. The control also maintains the roper ratio of fuel and air at all rates of combustion.

To change the steaming rate, the automatic combustion control must adjust the following:

1. Fuel oil to the furnace
2. Combustion air to the furnace
3. Feedwater to the steam drum

The Marine Combustion Control System consists of three regulators mounted in a single cabinet and three motor driven actuators mounted on top of this cabinet. The steam pressure regulator controls loading of fuel oil pressure regulating valves and loading of air volume regulators in response to steam pressure variations in the superheater outlet header.

The fuel oil pressure regulating valves automatically regulates fuel oil pressure to burners in direct proportion to the spring loading which is controlled by the steam pressure regulator. Therefore, the position of the steam pressure regulator is directly proportional to fuel oil pressure at the burners and is approximately proportional to quantity of fuel oil being burned. The air volume regulators control air flow through the burner registers in response to mechanical loading from steam pressure regulators. Ratio of fuel and air to each boiler is adjustable by means of a fuel air ratio knob and indicator for each boiler. The ratio adjusting mechanism is connected in the loading linkage between steam pressure regulator and air volume regulators.

Each air volume regulator maintains a balance between mechanical loading from the steam pressure regulator and air flow pressure differential, which is connected to the differential diaphragm of the air volume regulator. The air flow pressure differential is measured across the burner register, hence, no adjustment should be required when burners are taken out or put into service. When a burner is taken out of service, fuel oil is shut off and the air register is closed. Therefore, total air flow to the furnace is proportional to the number of burners in service without changing the relationship between the air flow pressure differential and fuel oil pressure at the burners.

Each regulator is equipped with a separate actuator consisting of a reversible motor which drives the power arm through a double worm gear reducing unit. A handwheel and a disengaging clutch are mounted on the secondary worm shaft at the front of the unit to provide the operator with a self-locking manual control which is instantly available in case of power failure. A mechanical position indicator is located adjacent to each manual control handwheel to indicate the position of the power arm of each regulator and the equipment being controlled.

STEAM PRESSURE REGULATOR

The steam pressure regulator consists of a bourdon tube pressure element, primary contacts, primary relays, stabilizer, and motor driven actuator. The steam pressure regulator serves as a master controller which varies the loading of fuel oil pressure regulating valves and air volume regulators in response to steam pressure variations.

STEAM PRESSURE REGULATOR OPERATION

The bourdon tube steam pressure element actuates the primary contact finger C between increase and decrease contacts C1 and C2 in response to steam pressure variations in the superheater outlet header. The bourdon tub tip carries the mounting plate for contacts C1 and C2 and the breakaway mechanism which connects with the right hand end of amplifying arm H. Amplifying arm H pivots about the lower end of the link supported by adjusting screw 0, and the left end of arm H connects to center contact finger C. When steam pressure increases, the tip of the bourdon tube moves downward, causing primary contact finger C to make contact with C2 which energizes the decrease primary relay. Contacts of the decrease primary relay energize the decrease coil of the reversing starter through the upper limit switch. Energizing the decrease coil of the reversing starter causes the motor driven actuator to drive the power arm downward to reduce steam pressure.

The action of the stabilizer to prevent over travel for the above operation is as follows:

1. Arm Y, which is keyed to the power arm shaft, is connected to stabilizer arm D through links U and N. hence, as power arm moves down to reduce steam pressure, link U moves down and link N moves up, pulling right hand end of stabilizer arm D upward. Arm D pivots about point E; hence point P, which supports stabilizer bell arm T, also moves upward. Arm T momentarily pivots about point F at the top of the stabilizer bell, and the right hand end of arm T moves up, compressing spring K. The actuator continues to run in the decrease direction until compression in spring K forces the tip of the bourdon tube upward, causing contact C and C2 to break and stop the actuator. Hence, motion of actuator is proportional to deviation in steam pressure. From this description of the operation, it is obvious that the primary function of the stabilizer and spring K is to proportion the motion of the actuator to the deviation in steam pressure.
2. Secondary function of the stabilizer is to release compression in spring K to the original value in the same length of time required for steam pressure to drop to the original control point, as a result of the motion of actuator power arm. This reset action is accomplished because the right hand end of arm T is restored to the original position by springs R and W. During this period of reset, arm T pivots about point P. The rate at which the right hand end of arm T is restored depends on the rate at which oil flows from the inside of the stabilizer bell through valve V.
3. Operation of the regulator for a decrease in steam pressure is exactly the same as explained above, except the motions are all in the opposite direction and spring K is extended instead of being compressed.
4. To obtain stable operation of the regulator, it is necessary to adjust the number of active turns in spring K so that the motion of the actuator power arm delivers the required loading to fuel oil pressure regulating valves and air volume regulators for a given deviation in steam pressure. Then valve V must be adjusted so that the stabilizer releases the tension or compression in spring K to the original value in the same length of time that is required for steam pressure to be restored to the control point.
5. The bourdon tube pressure element is provided with a loading spring L for adjusting steam pressure control point. To increase steam pressure, increase tension in loading spring L.

6. The 1/2" pipe from superheated steam header to steam pressure regulator should slope toward the regulator so that this line will fill with condensate from the regulator to the header. A valve should be provided in this line near the control cabinet. If superheater outlets are not connected to a common header in the boiler room, it is necessary to use a 1/2" tie line between the two headers with a valve at each header. This tie line should be at a lower elevation than the headers so that the tie line will remain full of condensate. Connection to the regulator should be taken from a tee in this tie line so the regulator can be operated from either header. The valve at one end of the tie line must be kept shut when both boilers are operating so that across flow will not force the condensate from the tie line.
7. *CAUTION:* Shut valves in steam pressure lines connecting bourdon tube pressure element to steam headers, and vent pressure from bourdon tube before hydrostatically testing headers at pressure in excess of normal operating pressure to prevent damage to pressure element.

8. An Off-On power switch for regulator control circuits is located on the front of control cabinet under actuator handwheel.

AIR VOLUME REGULATOR

The air volume regulator consists of a mechanically loaded differential pressure diagram element primary contacts, primary relays, stabilizer, and motor driven actuator. The air volume regulator controls air flow through burner registers to maintain a balance between mechanical loading and air flow differential pressure on the differential diaphragm.

AIR VOLUME REGULATOR OPERATION

The differential diaphragm element is connected to balance beam H, which pivots about point 0, and the left end of beam H is connected to loading springs which are operated by loading cam. The loading cam is operated by the steam pressure regulator.

Contact finger C is connected to balance beam H so that any imbalance of the beam causes contact C to make with contact C_1 and C_2, depending on direction of imbalance. Hence, when loading force on left hand end of beam H is increased, contact finger C makes the contact C_2, and air volume regulator decreases air flow so that air flow differential pressure applied to the diaphragm balances decreased loading force. When contact finger C makes with contact C_2, the decrease primary relay is energized. Contacts of the decrease primary relay energize the decrease coil of reversing starter through upper limit switch. Energizing the decrease coil of the reversing starter causes motor driven actuator to drive power arm down to reduce air flow.

Action of the stabilizer to prevent over travel for the above operation is as follows:

1. Arm Y, which is keyed to power arm shaft, is connected to stabilizer arm D through links U and N. Hence, as power arm moves down to reduce air flow, link U moves down and link N moves up, putting the right hand end of stabilizer arm D upward. Arm D pivots about point E; hence, point P, which supports stabilizer bell arm T, also moves up. Arm T momentarily pivots about point F at the top of the stabilizer bell, and the right hand end of arm T moves up, compressing spring K which opposes the decrease loading force on beam H and tends to raise contact finger C. The right hand end of stabilizer bell arm T is also connected through linkage to contact yoke CY, on which contacts C_1 and C_2 are mounted. When the right hand end of arm T moves up, contact C_2 is moved down, tending to break contact with finger C. The actuator continues to run in the decrease direction until the compression of spring K, tending to raise contact finger C, and the downward motion of contact C_2 causes the contacts to break and stop the actuator. From this description of the operation, it is obvious that the primary function of the stabilizer is to proportion the motion of the actuator to the unbalanced force on beam H produced by a change in loading force or a sudden deviation in air flow
2. The secondary function of the stabilizer is to restore (reset) the right hand end of stabilizer bell arm to the original position at an adjustable rate after each actuator motion. Thus, compression in spring K and the position of contacts C_1 and C_2 are restored to the original position as forces on beam H are rebalanced. Spring W, which is attached to bell arm T, balances the weight of the bell so that the pin in the right hand end of arm T is always restored to the position shown between the centering fingers of exactor assembly R. The purpose of exactor assembly R is to overcome friction in the linkage and exactly position the pin in the end of arm T. The rate at which the right hand end of arm T is restored to the original position depends upon the rate at which oil flows through valve V in stabilizer bell.
3. Operation of the regulator for an increase in loading force on beam H is exactly the same as described above, except that the motions are all in the opposite direction; hence, contact finger C makes with C_1 and spring K is extended instead of compressed.
4. To obtain stable operation of the regulator, it is necessary to adjust the number of active turns in spring K so that motion of the actuator power arm makes the required change in air flow for a given change in loading force on balance beam H. Then valve V in the stabilizer

bell must be adjusted so that the stabilizer releases (resets) tension or compression in spring K to the original value and restores contacts C1 and C2 to the original position in a slightly longer time then required for air flow differential pressure to rebalance beam H.

5. Diaphragm connections inside the control cabinet are made with 1/8" copper tubing to dampen pulsations. The 1/2" pipe lines from the control cabinet to windbox and furnace pressure connections must be kept open by periodically disconnecting lines from the diaphragm chamber and blowing compressed air back through the lines to windbox and furnace pressure connections.
6. *CAUTION:* When blowing out draft lines to windbox and furnace pressure connections, disconnect tubing to the diaphragm chamber so compressed air cannot enter chamber and rupture diaphragm.
7. An off-on power switch for regulator control circuits is located on front of the control cabinet under actuator handwheel.

FUEL-AIR RATIO LINKAGE

The fuel oil ratio linkage connects power arm of the steam pressure regulator with a loading cam for each air volume regulator, and provides an adjustable ratio arm for each air volume regulator to guide loading cams at an angle to produce the desired ratio of air to fuel in each boiler.

The power arm of the steam pressure regulator, which is connected to loading arms of fuel oil pressure regulating valves by ball and socket links BL, is also connected by ball and socket link X to the bell crank which drives shuttle bar M. Loading cams for air volume regulators are connected to each end of the shuttle bar by pantograph linkage so that the cams are free to move in horizontal or vertical planes, but slope of the lower edge of the cam always remains constant in relation to the base of the control cabinet, regardless of the position of the cam in vertical or horizontal planes.

The loading cam of each air volume regulator positions the cam follower arm in proportion to the position of the power arm of the steam pressure regulator, which also positions loading arms of fuel oil pressure regulating valves. Ratio of motion of the cam follower arm to motion of the power arm of the steam pressure regulator is determined by position of the ratio arm. The loading cam is provided with a ratio arm follower which rolls along the lower edge of the ratio arm. Position of the ratio arm of each air volume regulator is adjusted by a separate fuel air ratio adjusting knob on top of the control cabinet. The adjusting knob positions the ratio arm through a self locking worm and gear. Each worm gear shaft also drives a fuel air ration indicator on front of the control cabinet.

A cam follower spring is hooked in the end of the cam follower arm to keep the loading cam follower in contact with the cam and to keep the ratio arm follower in contact with the ratio arm. The linear loading spring is connected to a bracket on the cam follower arm and by a wire link to beam H of the differential diaphragm element. Downward pull of the loading spring on beam varies linearly with the position of the cam follower arm. The air volume regulator controls air flow to maintain a balance between air flow differential pressure applied to the differential diaphragm and pull of the loading spring.

F.O. PRESSURE REGULATING VALVES

A fuel oil pressure regulating valve is supplied for each boiler. This valve is an automatic pressure regulating valve actuated by an internal bellows. The bellows receives its power from fuel oil pressure, therefore no external source of power is required. This valve regulates fuel oil pressure to burners in proportion to the position of the loading arm. Each valve can be adjusted so that the minimum pressure for which it will control is sufficiently high to prevent loss of ignition. The valve is equipped with compensating springs which compensate for the characteristic of the loading springs, so that fuel oil pressure to burners will be maintained at a constant value corresponding to the position of the loading arm, regardless of quantity of fuel oil being burned or number of burners in service.

COMBUSTION CONTROL SYSTEM OPERATION

When putting boilers into service, manually control fuel oil pressure by the handwheel of the steam pressure regulator and manually control air to each boiler by the handwheels of the air volume regulators. When steam pressure has been raised to operating pressure, approximately 600 psi, the steam pressure regulator can be put into service. To put regulator into service, turn Power switch (under handwheel) for regulator control circuits to ON position and allow about a minute for the primary relay vacuum tube temperature to rise.

If air flow differential pressure on the diaphragm does not exactly balance loading from the steam pressure regulator, the clutch hub which is driven by the actuator motor will begin to turn. If clutch hub turns in a clockwise direction, this indicated that loading is greater than air flow differential pressure. Turn fuel air ratio knob in the direction to move the fuel air ratio indicator to the left until regulator is balanced and clutch hub will stop turning. If clutch hub turns in counterclockwise direction, fuel oil ratio indicator should be moved to the right until clutch hub stops turning. When clutch hub stops turning, this indicates that balance beam is in balance and that primary contacts are open. The clutch of the actuator can then be engaged by pushing the handwheel inward and the regulator is in automatic operation. When air volume regulators are in automatic operation, fuel air ratio can be adjusted to correct the CO_2 for varying burner conditions and for different size burner tips, by turning the fuel air ratio knob.

The fuel air ratio indicator is moved to the right to increase air and to the left to decrease air. Any one or all of the regulators can be taken out of service by turning the power switch for regulator control circuits to the OFF position. Then manual control is available by pulling out on the handwheel to disengage the clutch. When variable speed forced draft fans are used, fans should be transferred to high speed when air volume regulators approach the wide open position, and fans may be transferred back to low speed when boiler load is reduced to the point where low fan speed will provide the required air for proper combustion. If inlet vanes are used on forced draft fans, vane linkage must be properly maintained to prevent development of lost motion.

BOILER ESSENTIAL CONTROL

A boiler is used to heat feed water in order to produce steam. The energy released by the burning fuel in the boiler furnace is stored (as temperature and pressure) in the steam produced. The essential requirement for a combustion control system is to correctly proportion the quantities of air and fuel being burnt. This will ensure complete combustion, a minimum of excess air and acceptable exhaust gases. The control system must therefore measure the flow rates of fuel oil and air in order to correctly regulate their proportions.

A combustion control system capable of accepting rapid load changes is shown in Figure. Two control elements are used, 'steam flow' and 'steam pressure'. The steam pressure signal is fed to a two-term controller and is compared with the desired value. Any deviation results in a signal to the summing relay. The steam flow signal is also fed into the summing relay. The summing relay which may add or subtract the input signals provides an output which represents the fuel input requirements of the boiler. This output becomes a variable desired value signal to the two-term controllers in the fuel control and combustion air control loops.

A high or low signal selector is present to ensure that when a load change occurs the combustion air flow is always in excess of the fuel requirements. This prevents poor combustion and black smokey exhaust gases. If the master signal is for an increase in steam flow, then when it is fed to the low signal selector it is blocked since it is the higher input value.

When the master signal is input to the high signal selector it passes through as the higher input. This master signal now acts as a variable desired value for the combustion air sub-loop and brings about an increased air flow. When the increased air flow is established its measured value is now the higher input to the low signal selector. The master signal will now pass through to bring about the increased fuel supply to the boiler via the fuel supply sub-loop. The air supply for an increase in load is therefore established before the increase in fuel supply occurs. The required air to fuel ratio is set in the ratio relay in the air flow signal lines.

BOILERS AND BOILER CONTROL SYSTEMS (ENERGY ENGINEERING)

A boiler is a closed vessel intended to heat water and produce hot water or steam through the combustion of a fuel or through the action of electrodes or electric resistance elements. Many commercial and industrial facilities use boilers to produce steam or hot water for space heating or for process heating. Boilers are typically major users of energy, and any person involved in a facility's energy management needs to know how a boiler works and how the performance of a boiler can be maintained or improved.

In particular, it is important to know what parameters of a boiler system are the most important. For fossil fuel-fired boilers, the combustion efficiency is the major parameter of interest; this is most often controlled by providing the optimum amount of combustion air that is mixed with the fuel. Thus, understanding boiler control systems is extremely important. Steam and hot water boilers are available in standard sizes from very small boilers for apartments and residences to very large boilers for commercial and industrial uses.

BOILER TYPES

Boilers are classified by water temperature or steam pressure. They are further classified by the type of metal used in construction (cast iron, steel, or copper), by the type of fuel or heat element (oil, gas, or electricity), or by the relationship of fire or water to the tubes (i.e., firetube or watertube).

- Low-pressure boilers are those designed to produce steam up to 15 psig or hot water up to 250°F with pressures up to 160 psig.
- Medium- and high-pressure boilers produce steam above 15 psig or hot water above 160 psig or 250°F or both.

Boilers are typically constructed of cast iron or welded steel. Cast iron boilers are made of individually cast sections and are joined together using screws or nuts and tie rods or threaded rivets. The number of sections can be varied to provide a range of capacities. Steel boilers come in a wide variety of configurations. They are factory-assembled and welded and shipped as a unit. The fire and flue gases are substantially surrounded by water. The products of combustion pass through tubes to the back then to the front and once more to the back before finally exiting at the front. This makes it a four-pass boiler.

Firetube boilers are manufactured in many other configurations such as:

- External firebox—The firebox is not surrounded by water.
- Dry back—Firetubes are directly available from clean-out doors at the back of boiler.
- Scotch-Marine—Employs low water volume and has a fast response.

Watertube boilers are steel body boilers used for high capacity requirements of more than 2 million Btu per hour (Btu/h). Watertube boilers use a water-cooled firebox which prolongs the life of furnace walls and refractories.

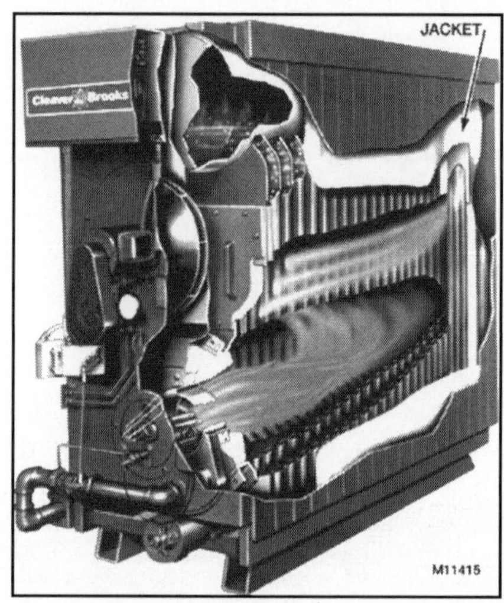

Fig. Typical cast iron boiler (watertube).

Modular boilers are small, hot water boilers rated from 200,000 to 900,000 Btu/h input. These boilers are available with gross efficiencies of 85% or higher. The features of a typical modular boiler. These boilers are often used in tandem to provide hot water for space heating and/or domestic hot water. For example, if the designed heating load were 2 million Btu/h, four 600,000 Btu/h (input) modular boilers might be used. If the load were 25% or less on a particular day, only one boiler would fire and cycle on and off to supply the load. The other three boilers would remain off with no water flow. This reduces the flue and jacket (covering of the boiler) heat losses.

Some modular boilers have a very small storage capacity and very rapid heat transfer so water flow must be proven before the burner is started. Electric boilers heat water or produce steam by converting electrical energy to heat using either resistance elements or electrodes. Electric boilers are considered to be 100% efficient since all power that is consumed directly produces hot water or steam. Heat losses through the jacket and insulation are negligible and there is no flue.[1]

Electrode boilers have electrodes immersed in the water. Electrical current passes through the water between electrodes, and this current and the resistance of the water results in generated heat. Electrode boilers are available in sizes up to 11,000 kW. Resistance boilers have the resistance (heating) elements immersed in, but electrically insulated from, the water and are manufactured in sizes up to 3,000 kW. Electric elements and electrodes are usually grouped to provide four or more stages of heating. A step controller responds to steam pressure or hot water temperature, activating each stage of heating as required to heat the building.

BOILER RATINGS AND EFFICIENCY

Boilers can be rated in several ways. The commonly used ratings and terms. The terms Btu/h (Btu per hour) and MBtu/h or MB/H (1000 Btu/h) indicate the boiler's input rate. Input ratings are usually shown on the boiler's (or burner's) nameplate. The terms bhp (boiler horse power), EDR (equivalent direct radiation), and pounds per hour (of steam) indicate the boiler's output rate.

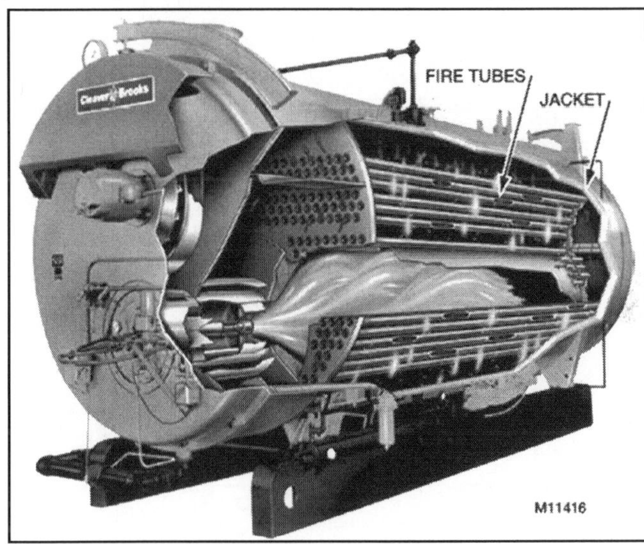

Fig. Typical firetube boiler.

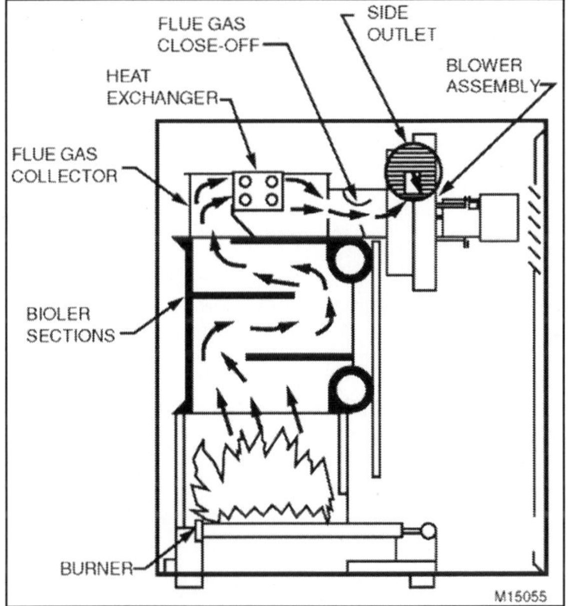

Fig. High efficiency modular boiler.

Gross efficiency of the boiler is the output (steam or water heat content and volume) divided by the fuel input (measured by a fuel meter at steady-state firing conditions). The combustion efficiency, as indicated by flue gas conditions, does not take into account jacket, piping, and other losses, so it is always higher than the gross efficiency.

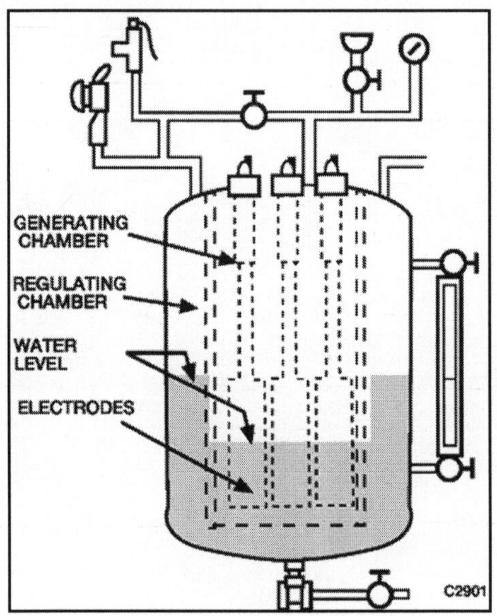

Fig. Electrode steam boiler.

A testing procedure issued by the U.S. Department of Energy in 1978 measures both on-cycle and off-cycle losses based on a laboratory procedure involving cyclic conditions. The result is called the AFUE (Annual Fuel Utilization Efficiency) rating, or seasonal efficiency, which is lower than gross efficiency.

COMBUSTION IN BOILERS

When gas, oil, or another fuel is burned, several factors must be considered if the burning process is to be safe, efficient, and not impact the environment.

The burning process must adhere to the following guidelines:

1. Provide enough air so that combustion is complete, and undesirable amounts of carbon monoxide or other pollutants are not generated.
2. Avoid excess air in the fuel-air mixture which would result in low efficiency.
3. Completely mix the air with fuel before introducing the mixture into the firebox.
4. Provide safety controls so that fuel is not introduced without the presence of an ignition flame or spark and so that flame is not introduced in the presence of unburned fuel.

5. Avoid water temperatures below the dewpoint of the flue gas to prevent condensation on the fireside of the boiler.

Combustion can be monitored by flue gas analysis. For large boilers, over 1,000,000 Btu/h, the analysis is typically continuous. For small boilers, flue gas is analyzed periodically using portable instruments. Flue gas composition analysis routinely measures the percentage of CO_2 (carbon dioxide) or 02 (oxygen), but generally not both. Ideal CO_2 concentration is in the 10%-12% range. The percentage of oxygen remaining is the most reliable indication of complete combustion. The ideal 02 concentration in the flue gas is in the 3%-5% range. Lower concentrations are impractical and often unsafe. Higher 02 concentrations mean that an excessive quantity of air is being admitted to the combustion chamber and must be heated by the fuel. This excess air passes through the boiler too quickly for the heat to be efficiently transferred to the water or steam, and thereby reduces the combustion efficiency. CO_2 measuring instruments are simpler and cost less than 02 measuring instruments.

The CO_2 or 02 concentration, plus the stack temperature, provides a burner combustion efficiency in percent— either directly or by means of charting. This combustion efficiency indicates only the amount of heat extracted from the fuel. It does not account for excess heating of combustion air, or losses from leaks or the boiler jacket, among other factors.

For oil-fired boilers, the oil burners are usually of the atomizing variety, that is, they provide a fine spray of oil.

Several types of these oil burners exist:

- Gun type burners spray oil into a swirling air supply.
- Horizontal, rotary burners use a spinning cup to whirl oil and air into the furnace.
- Steam- or air-atomizing burners use high pressured air or 25 psig steam to break up the oil into fine droplets.

For modulating or high/low flame control applications, the rotary or steam/air-atomizing burners are most common. For natural gas-fired boilers, the two typical types of gas burners are the atmospheric injection burner and the power type burner. The atmospheric injection burner uses a jet of gas to aspirate combustion air and is commonly used in home gas furnaces and boilers. The raw-gas ring burner is an atmospheric injection burner. Power burners use a forced-draft fan to thoroughly mix air and gas as they enter the furnace. Common power burner applications are in the commercial and industrial sectors.

BASIC BOILER CONTROLS

Boilers have to provide steam or hot water whenever heat is needed. A conventional BMCS (boiler management control system) is often set to provide a continuous hot water or steam supply between October and May at anytime the OA (outside air) temperature drops to 60°F for more than 30 min and an

AHU (air handling unit) is calling for heat. The BCMS should include a software on/ off/auto function. Unlike chillers, boilers can be left enabled at no-load conditions, during which time the water temperature will be held at the designed temperature. Frequent warm-up and shut-down of boilers causes stress buildup. Boiler manufacturers' recommendations provide specific guidelines in this area of operation.

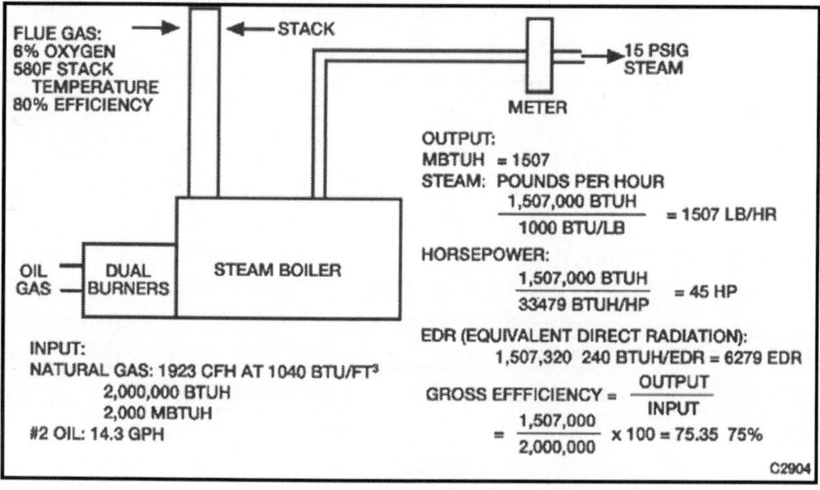

Fig. Boiler ratings and efficiency.

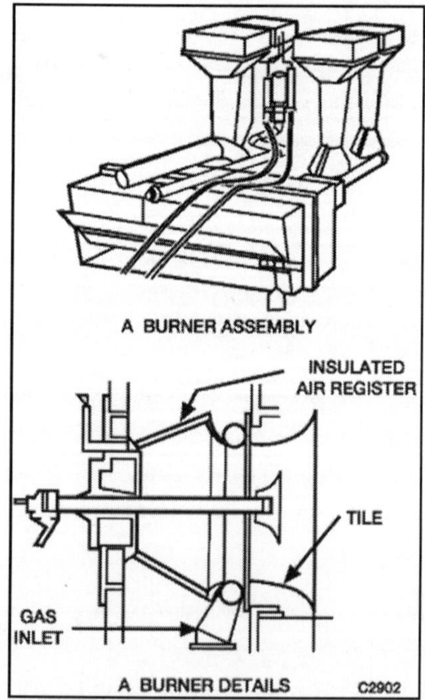

Fig. Raw gas ring burner.

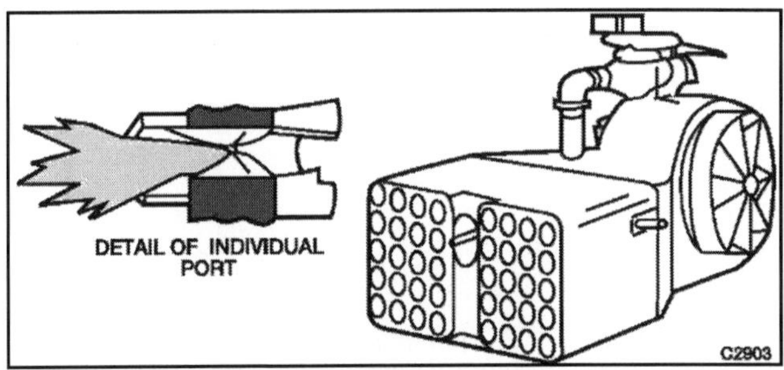

Fig. Multiport forced-draft gas burner.

Unless a low-limit for water temperature is used, hot water boiler burners are not controlled to provide water temperatures based on outdoor temperatures, because the reset schedules require water temperatures to be supplied below the dewpoint temperature of the flue gas. Some boilers require incoming water temperatures to be above 140°F before going to high-fire. In this case, if a building is using a hot water system and the boiler is locked into low-fire because the incoming water is too cold, the system may never recover.

The following are three ways to control the output of a commercial boiler:

1. On/off (cycling) control
2. High-fire/low-fire control
3. Modulating control

On/off (cycling) control is most common for small boilers up to 1,000,000 Btu/h capacity. The oil or gas burner cycles on and off to maintain steam pressure or water temperature. Cycling control causes losses in efficiency because of the cooling (which is necessary for safety) of the fireside surfaces by the natural draft from the stack during the off, pre-purge and post-purge cycles.

High-fire/low-fire burners provide fewer off-cycle losses since the burner shuts off only when loads are below the low-fire rate of fuel input. Modulating control is used on most large boilers because it adjusts the output to match the load whenever the load is greater than the low-fire limit, which is usually not less than 15% of the full load capacity. Steam pressure or hot water temperature is measured to determine the volume of gas or oil admitted to the burner.

Boiler firing and safety controls are boiler-manufacturer furnished and code approved. A BMCS usually enables a boiler to fire, provides a setpoint, controls pumps and blending valves, and monitors operation and alarms. Combustion control regulates the air supplied to a burner to maintain a high gross efficiency in the combustion process. More sophisticated systems use an oxygen sensor in the stack to control the amount of combustion air supplied. Smoke density detection devices can be used in the stack to limit the reduction of air so stack gases stay within smoke density limits. A continuous reading and/or recording of flue gas conditions—O_2 concentration percentage, stack temperature—is usually included in the control package of large boilers. A simple combustion control system contains a linkage that readjusts the air supply from the same modulating motor that adjusts the fuel supply. There may be a provision to stop the flow of air through the fluebox during the off-cycles.

FLAME SAFEGUARD CONTROL

Flame safeguard controls are required on all burners. Flame controls for large burners can be very complicated while controls for small burners such as a residential furnace are relatively simple. The controls must provide foolproof operation—i.e., they must make it difficult or impossible to override any of the safety features of the system. The controls also should be continuously self checked. For commercial and industrial burners, the flame safeguard control generally goes through a series of operations similar to the following.

- Purge the firebox of unburned fuel vapor (prepurge).
- Light the pilot.
- Verify that the pilot is lit.
- Open the main fuel valve.
- Verify that the flame is present as soon as fuel is introduced.
- Cut off the fuel supply promptly if the flame fails.
- Purge the firebox of any unburned fuel after each on-cycle (post-purge).

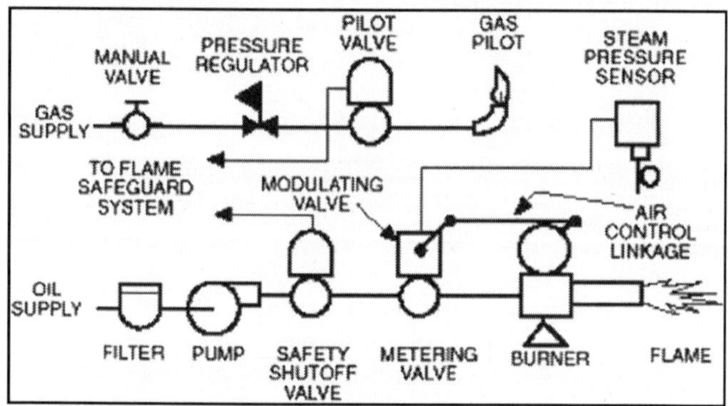

Fig. Combustion control for rotary oil burner.

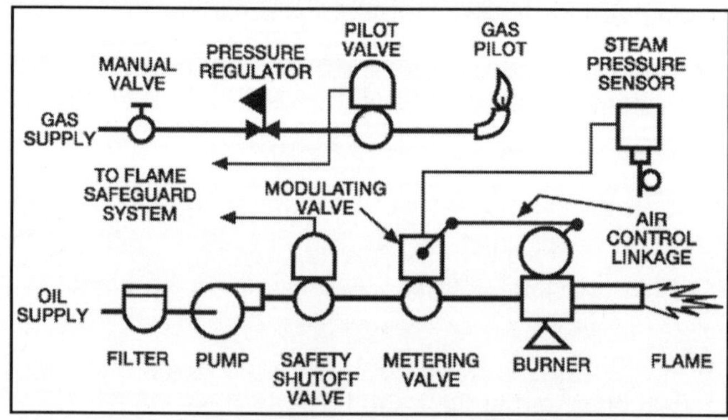

Fig. Simple flame safeguard for a gas furnace.

The key to any flame safeguard system is a reliable and fast means of detecting the presence or absence of a flame.

Methods of detection include:

- Response of a bimetal sensor to heat (slow response).
- Response of a thermocouple to heat (slow response).
- Flame conductivity (fast, but not reliable response)
- Flame rectification (fast, reliable response).
- Ultraviolet flame detection (fast, reliable response).
- Lead sulfide (photo) cells (fast, reliable response if a flame frequency check is included).

Some sensors can potentially malfunction because of short circuits, hot refractories, or external light sources. Other sensors, like flame rectification and ultraviolet detection, respond to flame only. Flame safeguard systems must be approved by Underwriter's Laboratory (UL) or Factory Mutual for specific applications. A flame safeguard system commonly applied to small gas boilers or furnaces. The flame of the gas pilot impinges on a thermocouple which supplies an electric current to keep the pilotstat gas valve open. If the pilot goes out or the thermocouple fails, the pilotstat valve closes or remains closed preventing gas flow to the main burner and pilot. The pilotstat must be manually reset.

How flame safeguard controls are integrated with combustion controls on a small, oil-fired steam boiler. The ultraviolet (UV) flame detector is located where it can see the flame and will shutdown the burner when no flame is present. In addition to the combustion, safety, and flame safeguard controls, larger burners often provide additional measuring instrumentation such as:

- Percentage of O_2 or CO_2 in flue gas (to monitor combustion efficiency)
- Flue gas temperature
- Furnace draft (in inches of water) column
- Steam flow with totalizer or hot water Btu with totalizer
- Oil and/or gas flow with totalizer
- Stack smoke density

CONTROL OF MULTIPLE BOILER SYSTEMS

Basic boiler connections for a three-zone hot water system. In this system, two boilers are connected in parallel. Hot water from the top of the boilers moves to the air separator which removes any entrapped air from the water. The expansion tank connected to the separator maintains the pressure in the system. The tank is about half full of water under normal operating conditions. Air pressure in the tank keeps the system pressurized and allows the water to expand and contract as the system water temperature varies. Water from the boiler moves through the separator to the three zone pumps, each of which is controlled by its own thermostat. In some systems, each zone may have a central pump and a valve. Return water from each zone goes back to the boiler in the return line. Several variations are possible within this type system, but the process is the same. There is no minimum boiler water flow limit in this example. The Dual Boiler Plant Control is a dual boiler plant with high-fire/low-fire controlled boilers. A minimum incoming water temperature of 145°F is required prior to high-fire, water flow must be maintained when the boiler is enabled, and a secondary hot water reset schedule of 110°F water at 55°F OA temperature and 180°F water at 5°F OA temperature. These concepts adapt well for single- or multiple-boiler systems.

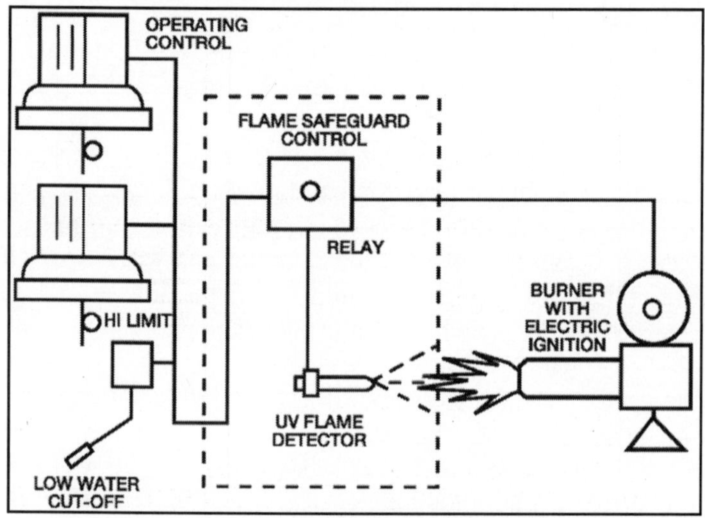

Fig. Combustion controls with flame safeguard circuit.

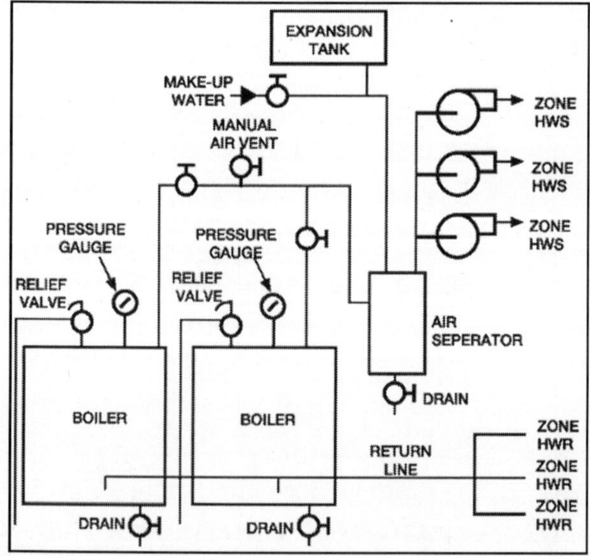

Fig. Typical piping for multiple-zone heating system.

Note: The primary/secondary decoupler is sized for the full secondary flow, and like the chiller plant decoupler, should be a minimum of 6 pipe diameters in length. Unlike the chiller decoupler, normal flow may occur in either direction.

Functional Description

Item no.	Function	Item no.	Function
1	On/off/auto function for secondary pumping system	7, 8 9	Off/auto function for boilers Heating system stop point (OA
2	On/off/auto function for heating system	10, 11	temperature) Operator information
3	Selects the lead boiler	12-14	Valve modulates to prevent
4	Heating system start point (OA temperature)		incoming water from dropping below the low-limit setpoint
5, 6	On/off/auto function for primary pumps		(145°F)

Item no.	Function
15-18	Secondary water setpoint reset from OA
19,20	Valve modulates to prevent incoming water from dropping below the low-limit setpoint (145° F)
21-23	Operator information
24	Icon, selects the Boiler System
	Control dynamic display
25,26	Software signal selection functions, allows valve to control secondary HW temperature, subject to boiler low-limits
27	OA reset valve control PID

Features

1. Full flow through operating boilers
2. Minimum temperature limit on the boiler's incoming water
3. Variable-flow secondary system with full boiler flow
4. Automatic boiler staging
5. User-friendly monitoring and adjustment

Conditions for Successful Operation

1. Control network, software, and programming to advise the heating plant controller of secondary fan and water flow demands.
2. Interlock and control wiring coordinated with the boiler manufacturer.
3. Control in accord with boiler manufacturer's recommendations.
4. Proper setpoint and parameter-project-specific settings.

Specification

The heating plant shall operate under automatic control anytime the secondary pump's on/off/auto function is not "OFF,"subject to a heating system's on/off/auto software function. The lead boiler, as determined by a software-driven lead-boiler-selection function, shall be enabled anytime the date is between October 1 and May 1, the OA temperature drops below 60°F for greater than 30 min, and an AHU is calling for heat. Each boiler's primary pump shall have a software on/off/auto function, and each boiler shall have a software auto/off function. The heating plant shall be disabled anytime the OA temperature rises to 65°F for greater than 1 min and after May 1.

BOILER SYSTEM CONTROL

A BLENDING VALVE ON EACH BOILER MODULATES IN THE RECIRCULATING POSITION TO PREVENT THE BOILER ENTERING WATER TEMPERATURE FROM DROPPING BELOW 145 DEGREES.

LEAD BOILER (1) AND ITS ASSOCIATED PUMP START ANYTIME THE OUTSIDE AIR TEMPERATURE DROPS TO 60 AND SHUTS DOWN ANYTIME THE OUTSIDE AIR TEMPERATURE RISES TO 65 DEGREES. ANYTIME THE LEAD BOILER STARTS FROM THIS OUTSIDE AIR TEMPERATURE SETTING, THE OTHER BOILER IS LOCKED OUT FOR 60 MINUTES.

ANYTIME THE LEAD BOILER CONTROL VALVE IS COMMANDED FULL OPEN BY THE SECONDARY WATER TEMPERATURE CONTROL LOOP FOR 5 MINUTES AND THE SECONDARY HOT WATER SUPPLY TEMPERATURE IS MORE THAN 5 DEGREES BELOW IT'S SETPOINT, THE LAG BOILER AND ITS ASSOCIATED PUMP START. ANYTIME BOTH BOILERS ARE OPERATING AND THEIR CONTROL VALVES ARE LESS THAN 40 PERCENT OPEN TO THE SECONDARY RETURN, THE BOILER SYSTEM OPERATING LONGEST SHUTS DOWN.

THE BOILER BLENDING VALVES MODULATE (SUBJECT TO THEIR LOW LIMIT CONTROL) TO PRODUCE SECONDARY WATER TEMPERATURES FROM 110 TO 180 DEGREES AS THE OUTSIDE AIR TEMPERATURE DROPS FROM 55 TO 5 DEGREES.

Fig. Boiler system control dynamic display.

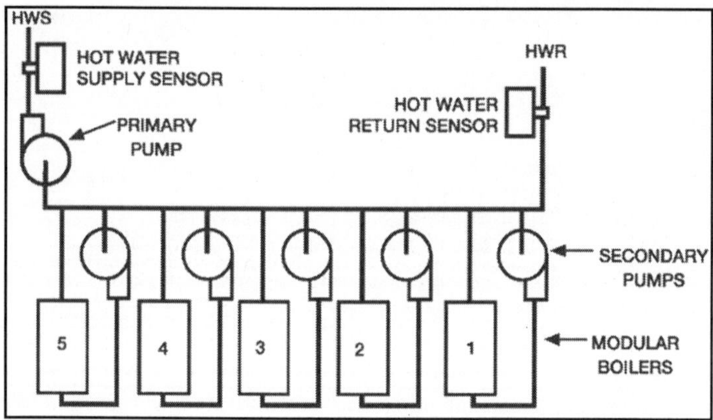

Fig. Typical primary-secondary piping for modular boilers.

Anytime the boiler plant is enabled, the lead boiler's primary pump shall start and, as flow is proven, the boiler shall fire under its factory controls to maintain 180°F. If the lead boiler's status does not change to "on," or if flow is not proven within 5 min, the lag boiler shall be enabled. During boiler operation, a three-way blending valve shall position to place the boiler flow in a recirculating mode until the water entering the boiler exceeds a low-limit value of 145°F, at which time the blending valve shall modulate to maintain the secondary water temperature between 110 and 180°F as the OA temperature varies from 55 to 5°F.

The lag boiler shall be locked out from operation for 60 min after the lead boiler starts. Thereafter, anytime one boiler control valve is commanded full open by the secondary temperature control loop for greater than 5 min and the secondary water temperature is a temperature less than 5°F below the secondary water temperature setpoint, the "off' (lag) boiler pump shall start. And, upon proving flow, the "off' boiler shall be enabled to fire under its factory controls to maintain 180°F. The just-started boiler's blending valve shall be controlled by an incoming 145°F water temperature low-limit sensor and setpoint similar to the lead boiler's, and subsequently, in unison with the other boiler's blending valve to maintain the reset, secondary hot-water temperature. Anytime both boilers are operating and their control valves are less than 40% open to the secondary return line, the boiler and pump that has run longest shall shut down.

MODULAR BOILERS

Modular boilers provide heat over a large range of loads and avoid standby and other losses associated with operating large boilers at small loads. A primary-secondary piping arrangement where each modular boiler has its own pump. The boiler pump is on when the boiler is on. Boilers that are off have no flow and are allowed to cool. Each boiler that is on operates at or near full capacity. Avoiding intermittent operation prevents losses up the stack or to the surrounding area when the boiler is off.

Normal control of modular boilers cycles one of the on-line boilers to maintain water temperature in the supply main to meet load requirements. The supply main's control sensor cycles the boilers in sequence. If the load increases beyond the capacity of the boilers that are on-line, an additional boiler is started. The lead (cycling) boiler can be rotated on a daily or weekly basis to equalize wear among all boilers or when using digital controls, the program can start the boiler that has been off the longest.

FEEDWATER REGULATOR

The ship is fitted with a Copes Feedwater Regulator System. The principal parts of the system are the type "P" thermostat with steam flow controller, the feedwater control valve, and the transfer and loading panel (control panel).

THERMOSTAT

The thermostat consists of an inclined stainless steel expansion tube mounted in a rigid channel steel frame. The upper end of the tube is connected to the steam space of the boiler drum end, the lower end to the water space. Water level in the tube always corresponds to that in the gauge glass. Water in the lower half of the expansion tube is cooler than steam in the upper half due to radiation. Thus, the tube cools and contracts as level rises. It heats and expands as level falls. Since upper end of the tube is fixed to the frame, any tube contraction or expansion rotates the thermostat lever. Outer end of the lever rises as water level rises. Lever motion is transferred through a bell crank lever to the stem of a pilot valve. This pilot varies air pressure in a control line in proportion to boiler water level. Control pressure increases as water level rises. The pilot is supplied with constant pressure air. It is pressure compensated so that it uses air only during water level changes.

STEAM FLOW CONTROLLER

Controlled air pressure from the pilot is piped to a Copes type "P" steam flow controller. This device includes a diaphragm type remote controller and a flow element. The remote controller is a spring loaded diaphragm which is connected to controlled pressure from the thermostat. The diaphragm position follows this pressure and thus is proportional to boiler water level. The flowmatic element provides a motion which is related to boiler steam flow and thus is proportional to boiler feed requirements. The flowmatic element consists of a spring loaded diaphragm in a pressure chamber. One side of the diaphragm is connected to boiler drum pressure and the other side to superheater outlet pressure.

Pressure loss through the superheater increases with boiler load. This pressure difference acts across the flowmatic diagram and is balanced by the spring. Thus, the diagram moves in response to changes in boiler load. Movement of the flowmatic element and remote controller are combined mechanically by external linkage. The combined movement positions the stem of a pressure compensated pilot valve. This pilot is supplied with constant pressure air. As the pilot stem is moved, the pilot changes air pressure in a control line. The control line is connected to a diaphragm receiver of the feedwater control valve.

FEEDWATER CONTROL VALVE

The control valve is a diaphragm type feedwater control valve. The diaphragm moves in response to air impulses received from the steam flow controller via the control panel. This valve is designed to shut as control pressure increases (lower boiler load or higher water level). For each change in control pressure, there is a proportional change in opening the feedwater control valve. Amount of feed is therefore controlled by both boiler steam flow and water level.

Effect of steam flow and water level can be predetermined by adjustments at the thermostat and steam flow controller. By proper adjustment of steam flow influence, the valve will be automatically positioned as required to feed the boiler at all loads. This results in constant water level at all loads. The thermostat serves merely as a check on the flowmatic element and to take care of minor discrepancies in water level.

The regulator can also be adjusted to hold automatically a lower water level at minimum boiler load than at maximum load. This type of regulation permits fuller utilization of drum space to absorb the swell in water level that occurs at a sudden increase in boiler load, and thus reduces the possibility of carryover.

Normal water setting (level that is automatically held by the regulator) can be changed by adjustments at the thermostat. The regulator permanently retains water level setting and type of control characteristics selected. Thus, no periodic readjustment is necessary. The flowmatic regulator responds instantly to sudden changes in boiler load and water level that occur during maneuvers. When load is suddenly increased, more steam bubbles are formed in the boiler tubes. Steam bubbles displace some water, forcing it into the boiler drum and causing a sudden rise in level known as "swell." Regardless of how feed is controlled, this swell requires appreciable time to boil away. The flowmatic regulator will restore the level to normal quickly. At the same time, the flowmatic eases the effect on the feed pumps of boiler load change. Conversely, when steam load is suddenly reduced, a "shrink" in water level results, and time is required to restore level to normal.

TRANSFER AND LOADING PANEL

The control panel provides for remote manual positioning of the control valves. Normally, Transfer Valve Pointer handle is at the Autoposition. Controlled pressure from the thermostat pilot merely passes through the transfer valve and on to the control valves. However, when Transfer Valve Pointer handle is thrown over to the Hand position, controlled pressure from the thermostat is cut off and the control valve is connected to the Hand Loading valve on the panel.

The hand loading valve is a reducing valve which is supplied with constant pressure air through a second reducing valve. As hand loading valve is opened, supply of air to control line is increased. This builds pressure in control lines. Controlled pressure can therefore be varied by changing the opening of the hand loading valve. This feature is useful when feeding manually, for example, during periods of raising steam.

Pressure gauges on the panel indicate automatic control pressure and manual control pressure at all times. This makes it possible to switch from automatic to hand control at any time with no disturbance to feed flow. It is merely necessary to bring hand loading pressure equal to automatic controlled pressure before the transfer valve is switched.

Ample adjustments are provided at the thermostat to adjust water level held by the regulator, and to widen or narrow water level variation over the boiler load range. Rough adjustment of the position of the thermostat lever is accomplished by the adjusting sleeve at upper end of the expansion tube. Final adjustment of water level is obtained by using the adjustable ball-point on the Slide Adjustment.

FORCED DRAFT BLOWER AND DUCTWORK

One single inlet, backward curved, multiblade, centrifugal type forced draft blower is provided for each boiler. Each blower normally delivers about 24,500 CFM of free air at 100° F. against approximately 13.6 inches of water static pressure. Each blower is capable of a maximum design capacity of 36,150 CFM of free air of 100° F. against 28.3 inches of water static pressure. Each blower is directly connected through a flexible coupling to a four speed electric motor of 200 H.P. at 1700 rpm.

Either of the two forced draft fans onboard are designed to carry both boilers steaming at low loads. Should either the port or starboard forced draft fan fails, you can continue to fire both boilers at a reduced steaming rate using the cross-over ductwork. Looking at the diagram should the port fan fail, you would close valve #2 and open valve #3 to divert the forced air from the starboard fan to both boilers. The starboard fan would be able to maintain both boilers at a low steaming rate while on high speed. Should the starboard fan fail, you would close valve #1 and open valve #3 to have the port forced draft fan maintain both boilers. If you were to operate only the port boiler using the starboard fan, you would close the registers on the starboard boiler, close valve #2 and open valve #3. At low steaming rates, you must also by-pass the air preheaters. This is done by opening the four dampers on each airheater.

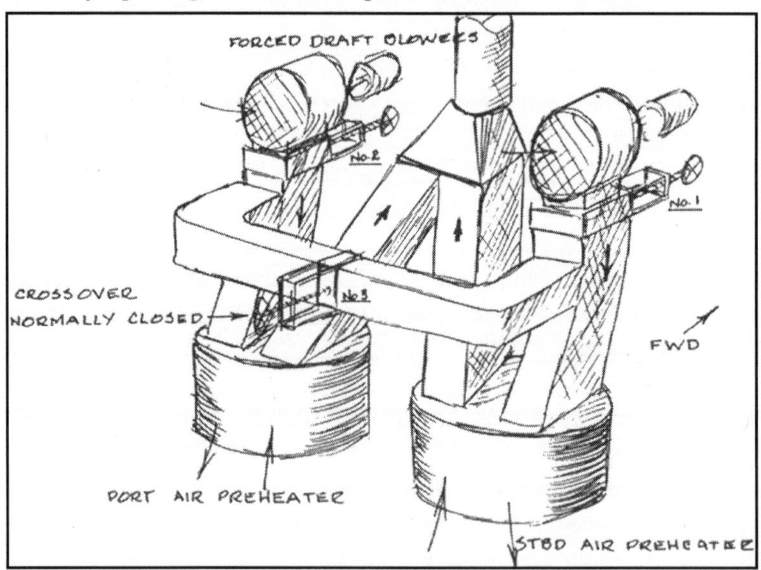

Fig. Forced Draft Blowers and Ductwork.

AIR HEATERS

Each boiler has one Ljungstrom vertical continuous counter flow regenerative type air preheater size 14, VMX-48 manufactured by the Air

Preheater Corporation. The regenerative type preheater consists of a series of parallel plates on a horizontally mounted wheel. The unit is divided into two sections of which the motor driven wheel (rotor) passes through both continuously at about 2 rev/min.

The cool incoming air from the forced draft fans pass through one side of the heater on the way to the double casing front to the oil burners. The hot exhaust gases pass up through the separate side of the heater on its way up the stack. The rotating plates are heated as they pass through the hot gases and liberate the heat as they rotate through the cool intake air. Radial seals located on the cold end of the air heater prevents the passage of hot exhaust gases to the air side and back down the double casing. Air heaters raise the efficiency of the plant 1% for every 40° F. rise in the incoming air temperature.

The cold end of the heater is exposed to severe heat conditions and sulfuric acid deposits. These plates are coated with porcelain to protect them from the corrosive conditions. The elements in the cold end are placed in buckets of Corten, installed and supported independently of the hot end layer. In order to remove the elements packed into the baskets for cleaning, special arrangements have been made for removal through the outer shell.

The drive unit consists of a low speed hydraulic unit with the motor connected directly to the rotor shaft. The Ljungstrom air preheater take up only 1/3 the volume (area) of a tubular type of the same capacity. Two soot blowers are provided for the removal of accumulated deposits along the heating surface. These electrically driven automatically controlled sootblowers are installed in the outlet area of the gas side and inlet area of the air side. Air leakage due to leaking radial seals is the principle source of trouble. These leaks would make the measured exit gas temperature appear lower than it actually is because of the cool incoming air being leaked up the stack. This apparently cooler exit temperature would give you a misleading preheater performance.

The air heater is provided with a by-pass so that when the boiler is operating at low firing rates, the uptake temperature can be maintained higher than the dew point of the uptake gases. This will prevent corrosive elements from the condensing vapors from settling on the heating surfaces of the air heater. As a general rule, the air preheater should be by-passed whenever the uptake temperature is less than 220° F. Also, in the event of a casualty to the air heater motor, the high exit gas temperatures would destroy the elements within the heater baskets. The air preheater has several points that should be checked regularly while in operation. These are, the rotor drive, the guide bearing and the support bearing. The rotor drive oil level should be checked every hour and the level kept at the line on the dipstick. Use care when checking this oil level as the dipstick can only be inserted to its lowest position when the red lines on the cap and pipe are lined up. The guide bearing oil level should be checked at least once each watch and the oil level kept at the line on the dipstick. The support bearing oil level should be checked every hour and the level kept between the lines on the dipstick. The temperature of the oil should be checked every hour and recorded. Should the temperature of the oil in the support bearing reach 160° F., the watch engineer should be notified at once. The pressure from the oil circulating pump should also be recorded. The oil used on the rotor drive, the guide bearing and the support bearing is CYLESTIC TK-680. The oil and oil can are stored in a metal box forward of the preheaters on the main deck level. The sight hole is provided to allow the watchstander a means of visual inspection of the cold end of the preheater. The watchstander should check for rotor rotation as well as the presence of any foreign matter or unusual noises.

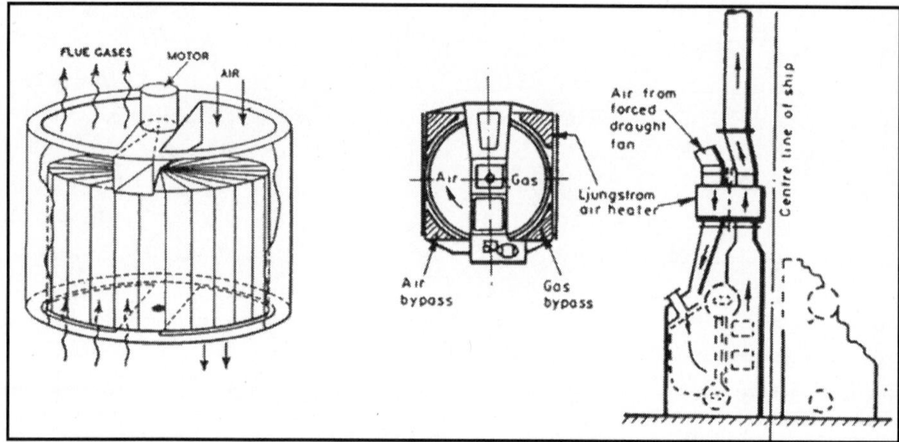

Fig. Ljungstrom Air Heater.

5

Marine Automatic Voltage Regulators

All modern ship alternators are equipped with voltage regulators. The regulators permit to obtain constant voltage of generators under various amount of load and power factor. Here we shall speak about AVR for brushless alternators. The regulator principle in an autonomous alternator system. The alternator voltage is fed to the AVR unit via fuses and exciter switch. The reference level of the alternator voltage is set by thc voltagc adjusting potentiometer. The difference between the actual voltage and reference value is amplified and converted to triggering pulses to the thyristor bridge. A-c voltage is fed from the alternator via supply transformer to the thyristor bridge.

Under normal circumstances, the voltage is kept constant. If speed drops of more than 5% occur, the AVR unit will automatically decrease the voltage proportional to the frequency (U/f-constant). To obtain correct load sharing between parallel running alternators, the alternator voltage is made current dependent. This is obtained by current transformer and a resistance network inside the AVR unit. A voltage build-up circuit which uses the alternators residual magnetism provides automatic voltage build-up.

The compounding circuit and a compounding transformer provide the necessary excitation current during short circuit conditions when voltage generator is equal to zero. It is necessary for immediate voltage build-up after elimination of short circuit. The main unit of regulator is AVR circuit. The AVR circuits consist of the following parts: voltage measuring circuit, amplifier, frequency measuring circuit, synchronizing and triggering circuit, stabilized power supply. All the above mentioned circuits are mounted on a printed board. The alternator voltage is transformed before it is fed to the printed circuit board via terminals. The voltage is rectified and a proportional voltage is fed to an operational amplifier where it is compared to the reference voltage. The difference voltage is amplified and controls, via a triggering circuit, the firing angle of the thyristors and thereby the exciter current. In case of alternator voltage change, the regulator generates a signal which changes the firing angle of the thyristors until the desired voltage is obtained. A stabilizing filter ensures stable and alternator voltage.

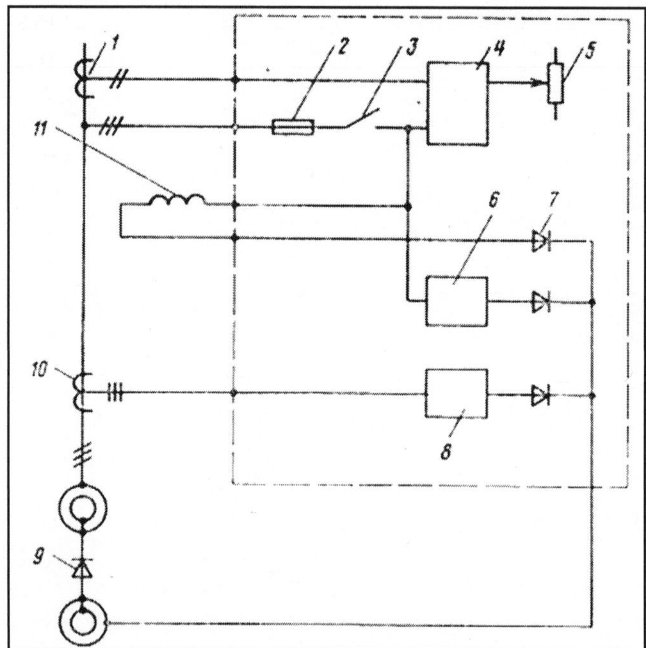

Fig. 1. Automate voltage regulator. Block diagram: 1 — current transformer; 2 — fuses; 3 — excitation switch; 4 — AVR unit: 5 — voltage adjuster; 6 — voltage build-up circuit; 7 — thyristor bridge; 8 — compounding transformer; 9 —rotating diode bridge; 10 — compounding circuit; 11 — supply transformer

STATIC AUTOMATIC VOLTAGE REGULATOR

The carbon pile regulator uses a magnetic coil powered from the alternator output Strength of the field vanes with alternator voltage and this strength is tested against springs which are the voltage reference. The moving contact regulator employs a similar matching of alternator output effect through a magnetic coil against springs.

The availability of a small transformed, rectified and smoothed power supply from the alternator output makes possible the matching or it directly against an electronic reference in the static automatic voltage regulator. The direct current derived from the alternator output is applied to a bridge which has fixed resistances on two arms and variable resistances (zener diode voltage references) on the other two.

The zeners operate in the reverse breakdown mode having been manufactured with a zener breakdown voltage of very low value. As can be seen from the earlier description of zener diodes, voltage remains constant once breakdown has occurred despite change in current. This implies, however, that changes in applied voltage while not affecting voltage across the diode will cause a change in resistance which permits change in current. .As with a Wheatstone bridge, imbalance of the resistances changes the flow pattern and produces in the voltage measuring bridge an error signal The error signal can be amplified and used to control alternator excitation in a number of different ways. Thus it can control the firing angle of Thyristors (Fig. 3.14) through a triggering circuit to give the desired voltage in the brushless alternator described. It can be used in the statically excited alternator to correct small errors through a magnetic amplifier arrangement-The error signal has also been amplified through transistors in series, for Excitation control.

SELECTING AN ALTERNATOR

Automotive-style alternators, like this68-amp Delco-style three-wire model from Sierra, are internally regulated.

LOOKING FOR AN ALTERNATOR?

We see customers looking for an alternator for two reasons here at West Marine: Direct replacement: you're needing a direct replacement for a factory-installed automotive-type OEM alternator, and we have a selection from Sierra alternators for most common marine engines. Go to our Engine Parts Selector to find the alternator you need.

Upgrade to your charging system: you want to replace your automobile-style alternator with a larger, higher-capacity alternator so you can charge more batteries, charge larger batteries or bring them up to charge more rapidly. In this case, you're looking for an alternator from Mastervolt or Balmar, and this type of system upgrade is the topic of this West Advisor.

WHAT TO CONSIDER BEFORE YOU BUY

Before you invest in a high-output charging system, save yourself a lot of trouble and dollars by ensuring that you buy the correct system for your application. Charging systems vary by type of boat, as well as a collection of other variables including engine type, engine room space, belt type and size, battery bank size and battery chemistry. Selection criteria you should consider include:

BATTERY CHEMISTRY

Each type of marine battery, a flooded starting battery, deep cycle flooded battery, gel or Absorbed Glass Mat (AGM) battery has its own unique charging characteristic, which affects the type and size of alternator you'll need to charge it efficiently. The Acceptance Rate of the battery, what percentage of its total amp hour capacity it can accept, is different for some battery chemistries.

Extra large frame 98-Seriesalternator delivers current similar to some gensets, with up to 5kW of output. A standard flooded battery, of either the starting or deep cycle type, can accept charging current equal to 25 percent of its available capacity. Gel batteries can accept 30 percent, AGM batteries can accept 40 to 50 percent, and newer technologies like lithium ion or TPPL (thin plate pure lead, such as Optima) can accept even more. They need to be charged by a large case or extra large case alternator to produce enough charging current, due to their nearly unlimited appetite for amps.

Your alternator should be capable of delivering rated output equal to the maximum acceptance rate of your house battery bank. In other words, the alternator's rating should be equal to 25 percent (lead acid) 30 percent (gel) or 45 percent (AGM) of your total battery capacity. For flooded batteries, size your alternator toward the bottom of this range.

BATTERY CAPACITY

As we add electrical and electronic appliances to our boats, we notice that our battery banks have become incapable of handling the loads created by these new stereo subwoofers, floodlights, satellite TV systems, refrigerators and other thirsty power consumers. We start shopping for bigger battery banks, or consider adding a battery. Adding battery capacity requires us to take a look at our charging system.

In most cases, your alternator's output should equal the maximum available capacity of your battery bank. In other words, if your house battery bank is made up of flooded batteries with an amp-hour capacity rating of 400Ah, the maximum acceptance rate of those batteries would be 25 percent, or 100 amps, the ideal alternator would also be rated at 100 amps. If your house bank consisted of 400Ah capacity AGM batteries, the maximum available capacity would be 45 percent of 400, or roughly 180 Ah. To maximize charging efficiency, a 180-amp rated alternator would be required. Unfortunately, mounting a large-case alternator is not feasible in many vessels.

MOUNTING STYLE

Compatibility depends on engine model, engine year, engine compartment layout and other factors. Inspect your existing alternator mount and compare to the diagrams shown to determine the appropriate high-output replacement. You may need spacers or modifications to brackets, tensioners and wiring on any high-output alternator installation. Most alternators fall into one of four mounting styles:

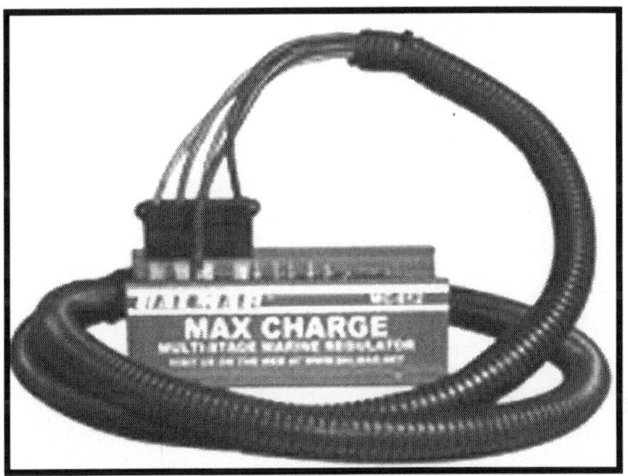

Max Charge 614-H has the latest fast processor, seven battery charging algorithms, and a Belt Load Manager program to control field output to ensure that belts are protected from excess wear.

- Single 1" foot spindle mount (Motorola style; Balmar Model 621). Typical engine types are Hino, Lehman, Caterpillar, Atomic 4, Universal, Ford, Crusader, Pathfinder, Westerbeke, other Motorola-equipped.
- Single 2" foot spindle mount (Delco style; Balmar Model 621 with included spacer).Typical engine types are Volvo, Cummins, Westerbeke, Perkins, MerCruiser, Yanmar (6LP), Volvo-Penta, GM, other Delco-equipped.
- 3.15" ID saddle mount (Hitachi style; Balmar Model 60). Typical engines are MerCruiser, Lehman, Yanmar, Westerbeke, Perkins-Sabre, other Hitachi-equipped.
- 4" ID saddle mount (J-180 style; Balmar Model 604). Typical engines are Detroit Diesel, Cummins, Caterpillar, John Deere, J-180 off-engine mounts.

BELT LOAD LIMITATIONS

Drive belt width is a critical factor in selecting a replacement alternator. Commonly-installed vee-belts and multi-groove serpentine belts have specific limits regarding the amperage and horsepower loads they can support. As a rule of thumb, 12-volt alternators use one horsepower for every 25 amps of output, or to put it another way, their ratio of output to horsepower load is 25:1. So, when selecting a new alternator, you need to size it for the limits of the capacity of its drive belt. Otherwise, your system will be plagued by premature belt wear, belt slippage and potential damage to your alternator and engine.

Belt Type	Belt Width	Max. HP Load	Highest Recommended Output
Single Vee	3/8	3.5	80A @ 12V, 30A @ 24V
Single Vee	1/2	4.5	110A @ 12V, 45A @ 24V
Dual Vee	1/2	12	310A @ 12V, 220A @ 24V
Serpentine	6-groove	NA	310A @ 12V, 220A @ 24V
Serpentine	8-groove	NA	310A @ 12V, 220A @ 24V

INTERNAL OR EXTERNAL REGULATORS

In automotive type alternators, current for the alternator is supplied by an internal regulator, which drives the alternator to a specific voltage value (usually about 14.1 volts), which works great for a starting battery, just like the one in your car. Deep cycle and sealed gel and AGM marine batteries require a more complex program of charging voltages to achieve their optimal charge. Multi-stage external voltage regulators, like the Balmar Max Charge and ARS-5 enable the alternator to vary charging voltages, based on the battery's temperature, chemistry (flooded, gel or AGM types) and level of discharge, to ensure that batteries are recharged quickly and safely.

COMMON UPGRADES FOR SOME TYPICAL BOAT TYPES

Sailboats: with a single propulsion engine doubling as the primary battery charging source, traditional monohull inboard sailboats require maximum charging output while operating under engine power. Replacing the stock alternator with a Balmar alternator and multi-stage regulator can substantially increase charging performance. Add the Digital Duo Charge for starting battery care.

- 6-Series High-Output Alternator
- ARS-5 Voltage Regulator
- Alternator Temperature Sensor
- Battery Temperature Sensor
- Digital Duo Charge

Multihulls: By combining the outputs of both port and starboard alternators, the multihull charging system can effectively charge twice as much battery capacity. Balmar's twin-engine Centerfielder II allows port and starboard systems to work independently or together, doubling the output. It monitors both port and starboard alternators and regulators, controls field output to both alternators, making it possible to direct the combined output from both alternators to charge a large central house battery bank.

- Two 6-Series High-Output Alternators
- Two MC-614 Voltage Regulators
- Centerfielder II
- Two Alternator Temperature Sensors
- Two Battery Temperature Sensors
- Two Digital Duo Charges

Twin-engine powerboats: as with multihulls, twin-engine powerboats can double their effective charging output with the Centerfielder II to control both port and starboard charging systems. Adding a Digital Duo Charge makes starting battery charging automatic. Common system upgrades:

- Two 6-Series High-Output Alternators
- Two MC-614 Voltage Regulators
- Centerfielder II
- Two Alternator Temperature Sensors
- Two Battery Temperature Sensors
- Two Digital Duo Charges

WHAT ELSE TO LOOK FOR

Safety: Your replacement alternator should meet U.S. Coast Guard Title 33 safety protection standards. If the alternator is being installed on a gasoline engine, the alternator should also comply with international SAE and ISO standards for ignition protection. Balmar alternators meet all three standards. Heavy-duty construction: marine alternators operate at sustained high output for much of their lives. This requires high-temperature grease, large cooling passages, large or multiple fans, heavy-duty bearings, high-amperage diodes, etc. Standard automotive-type alternators rarely offer the construction required to meet the demands of large marine charging systems.

Temperature compensation: The ambient temperature of your batteries and alternator both effect charging efficiency. Max Charge and ARS-5 regulators both sense battery temperature and can regulate charging voltage accordingly. Alternator temperature sensing can also save your alternator by reducing output if the alternator overheats. Both alternator and battery temperature sensing features require optional temperature sensors (MC-TS-A and MC-TS-B).

Wiring and fusing: Wiring between the alternator and the batteries must be matched to the amperage rating of the replacement alternator. A wiring chart, based on amperage output and wire length, is included in the Balmar alternator installation booklet.

HIGH POWER ALTERNATOR

A FEW BASIC FACTS ABOUT CONVENTIONAL ALTERNATORS

Fig. ZENA SR150B 150 modular power generator
150A @ 12-14.5VDC/ 80A @ 24-29VDC.

Fig. ZENA 200 amp. modular power generator
200A @ 12-14.5VDC/ 125A @ 24-29VDC.

A "standard" automotive or marine alternator is designed to produce a certain amount of power when the engine of the vehicle or vessel for which it is designed is idling. This is the amount of power that the vehicle needs for its basic onboard equipment — engine ignition/controls, fuel pumps, instrumentation, lights, climate control systems, etc. This type of alternator is usual identified by the maximum amount of power that it can produce when the engine is operating at high speed — however, it cannon produce this amount of power for more than a few minutes without overheating and potential damage. The most power that such an alternator can be expected to provide continuously is about 1/2 the maximum rated output power — possibly as much as 2/3 for a conventional "heavy duty" design. Special design alternators which can produce their full rated output continuously are available. The ZENA DC Power

Generator is this type of alternator — built to a standard of construction that far exceeds that of the standard alternator. Alternators made by companies like Leece-Neville and Niehoff are

Except in cases where only a little more charging current is required to handle, for example, a few extra lights for trailer towing, or a bit more output to help get a quicker charge during winter months (when a lot of your driving is done at slow engine speeds in icy/wet conditions), installing a high output alternator may be quite a bit more complicated that it was in the 1960's, 1970's, or 1980's when most American cars were equipped with a single size alternator with a case that was large enough so that it could be easily rebuilt, or replaced, with an alternator that looked identical, but produced as much as double the current provided by the vehicle's original alternator. Modern alternators are a much different prospect. Typically, for a given application (vehicle/engine), the alternator output is designed to exactly meet the vehicle's electrical needs when operating at low engine speeds and the alternator is built to be as physically small and light as possible (often making cooling under high load conditions very difficult) to help the manufacturer achieve better vehicle fuel economy. Often, with these alternators, power produced at low engine speeds is kept relatively low (or even turned off) to reduce the load on the engine, again to achieve better fuel economy.

With the addition of computer controls in vehicles and on vehicle engines, alternators also were changed to act as protective devices for the installed computer systems — further limiting their maximum charging capacity and making it even more difficult to rebuild or replace them with non-stock higher amp units. Also, in many cases, alternators also became active participants in the engine's computerized electrical controls systems. In fact, their interaction with the vehicle electrical system is so critical that, in most cases, replacement of a stock alternator with one that is not supplied by the vehicle/engine manufacturer will void the vehicle warranty and may also make the vehicle impossible to service with factory diagnostic equipment!

Making things even more complicated are newer sealed batteries such as gel cells and AGM (absorbed glass mat) — batteries (typically, after market components, that require different charging voltages and techniques than those used for the common lead acid battery. These batteries are different enough so that simply replacing an existing lead acid battery with a new sealed type battery in a vehicle equipped with a stock alternator with a built in, nonadjustable voltage regulator may require an alternator change just to protect the new battery from too high of a charging voltage!

But you STILL need a high amp charging system to power that super stereo system you installed, or you need to provide power for emergency equipment, or to drive a huge CB radio transmitter, or to power that a huge winch that you installed, or to supply current to an inverter producing on board AC power for tools and equipment, to drive an air conditioner on your RV or trailer, or even power your RV, or trailer's complete, electrical system.

INSTALLING A HIGH POWER MARINE ALTERNATOR ON YOUR BOAT.

Arguably, there is no more worthwhile electrical project you can undertake on your boat to improve comfort and convenience than the installation of a high power alternator. Having a high power alternator will provide you with a substantial improvement in the availability of electrical power on board. This will lead to greater safety, better ability to run electronics, better lighting, colder beer and the ability to run a watermaker, windlass, inverter and other high consumption devices. It is not the only component that matters in a fully functional marine electrical system, other parts will be addressed in later articles, but in many cases the alternator serves as the foundation of a modern high performance system.

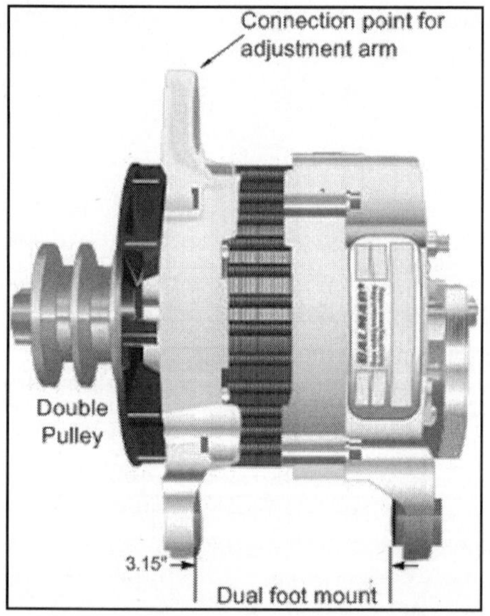

This drawing courtesy of Balmar shows a typical high output large case alternator. This model has a dual foot mount and a dual pulley for ½" belts and an external fan. A typical small to medium marine diesel has a standard alternator with a rating of 55 amps; however this rating is obtained under the most favorable conditions. The typical sustained output from these stock alternators is 25 – 30 Amps. It is easy to see how the addition of a 100 Amp high output alternator which has an actual output of 100 Amps could more than triple the amount of power available on board and substantially reduce engine running time.

Fortunately installing a replacement alternator is not difficult. It does involve some basic electrical and mechanical skills. I have attempted here to give quite detailed instructions and to cover all of the bases making the text that follows a bit long, but at the end of the day the job is not that difficult. To get started you first need to identify what kind of alternator you currently have.

CASE SIZE

Most smaller 4 and 6 cylinder diesels have what is known as a "small case alternator". The diameter of these alternators is about 5 ½" and the length including the pulley is about 6". They often have a single pulley for one belt although the more powerful ones will have double pulleys or a serpentine belt. Bigger engines normally have "large case alternators" which tend to be about 7" in diameter and about 7" long and having a double pulley or pulley for a serpentine belt. There are also "extra large case alternators" available for the biggest engines. The photo below shows the typical small case alternator found on a Yanmar

Fig. Mounting type.

The next think to observe is the mounting type. You need to see how the alternator is attached to the engine at its pivot point. Does the alternator have a single foot or does it have two?

DUAL FOOT ALTERNATORS:

Yanmar engines have alternators with two feet, known as a "dual foot mount". The standard spacing between the feet is 3.15", this is also known as a "Yanmar mount". This standard dual foot mount is also often found on Mercruiser and Mitsubishi engines as well as others. Many Westerbeke engines have alternators with two feet that are spaced closer together; the spacing between the feet is a bit less than 2". Larger alternators may have what is known as a J-180 mount with dual feet that are 4" apart. These are often used on larger Cummins, Detroit and Caterpillar engines.

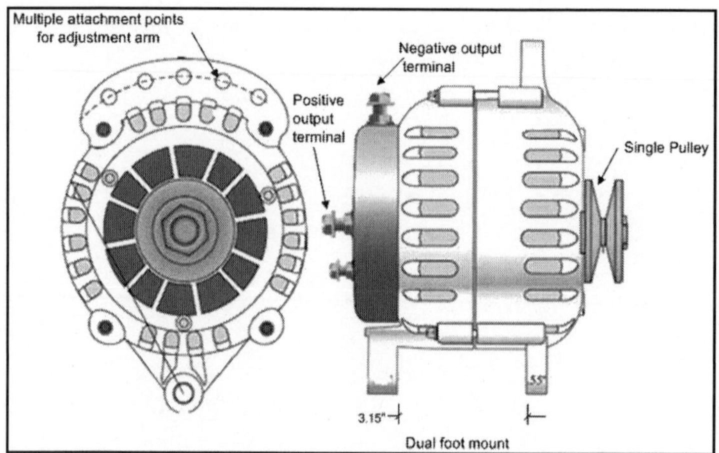

Fig. A small case dual foot alternator for use on a Yanmar or similar engine. The spacing between the feet is 3.15", This model has a single pulley and has multiple places to attach the adjustment arm. It has an internal cooling fan. Drawing courtesy of Balmar. Take a look at my Blog Post Installing a Balmar alternator on a Yanmar 3JH3E for a description and photos of an actual project.

SINGLE FOOT ALTERNATORS:

Single foot alternators fall into two categories, 1" foot and 2" foot mounts. The 1" or 2" refers to the dimension of the foot itself measured fore and aft. You will just have to look and measure but in general the 1" foot is used on some Caterpillar engines as well as Universal, Crusader, Atomic 4 and some Westerbeke's. The 2" foot is found on many Volvo's, Perkins and Lehman engines. This photo courtesy of Balmar shows a small case alternator with a single pulley and a single 2" mounting foot. There is no way to be sure of the mount without looking carefully at what you already have, there are just too many variables out there.

MOUNTING BOLT

The next thing to look at is the mounting bolt that attaches the alternator to the engine at its pivot point. The larger alternators have ½" diameter mounting bolts while most intermediate alternators have a 3/8" or 10 mm diameter mounting bolt (Note that these diameters are almost identical). The smallest alternators have a 5/16" or 8 mm mounting bolt. (Again these diameters are almost identical) It is important to get an alternator with the right size hole for the mounting bolt otherwise it wont fit or will wobble, it is usually possible to order a bushing to adapt to another size bolt, don't forget to ask when ordering.

PULLEY

Finally you need to look at the pulley: If your alternator has a single pulley you need to know the width of the belt. This is measured across the top of the belt, the widest part. Most single pulley alternators have a ½" belt, smaller ones have a 3/8" belt, and you will occasionally find some wider single belts. You may also find double belts or a serpentine belt. These are designed to transfer greater loads and have greater surface area in contact with the pulleys. If you have a serpentine pulley you need to count the number of grooves in the pulley and the overall width of the belt. The picture to the right shows an alternator with a single 1/2" drive belt, it also shows the alternator pulley perfectly in alignment with the engine pulleys. If you have a single 1/2" drive belt like this you are limited to an alternator of around 100 Amp output and if the belt is 3/8" diameter you are limited to an alternator of about 70 Amps output. These are not hard and fast rules and occasionally I have installed a single belt alternator of 120 Amps with a single 1/2" belt but to do that you have to be prepared to monitor wear on the belt more closely and be prepared to replace the belt more frequently. What about if you need more output than 100 Amps and are stuck with a single pulley system? One option is to use the Balmar Serpentine Conversion kit to change your engine over to serpentine pulleys. This clever arrangement allows you to fit new pulleys over the top of your existing crankshaft and water pump pulleys allowing you to install a much larger alternator. When used in conjunction with the Balmar AT-Series Alternator it allows you to install a 200 amp small case alternator on many small boat engines

OTHER MOUNTING CONSIDERATIONS

Occasionally there are other things to look at. Is there sufficient room behind the alternator in case the replacement is slightly longer? Will the pulley line up with the old pulley? Will the adjustment arm fit the way the old one did? These are snags that don't normally occur and have to be dealt with on a case by case basis. Before purchasing a replacement alternator it is worth discussing any unusual or unknown aspects of the installation with a professional. The photo below shows the Balmar Belt Buddy adjustment arm with tensioning device.

ELECTRICAL CONSIDERATIONS

All of the remaining part of this article refers to "P" type alternators, where a positive voltage is applied to the alternator field coils. It is possible to have the opposite, an "N" type alternator, with a negative connection to the field coils, but they are extremely rare.

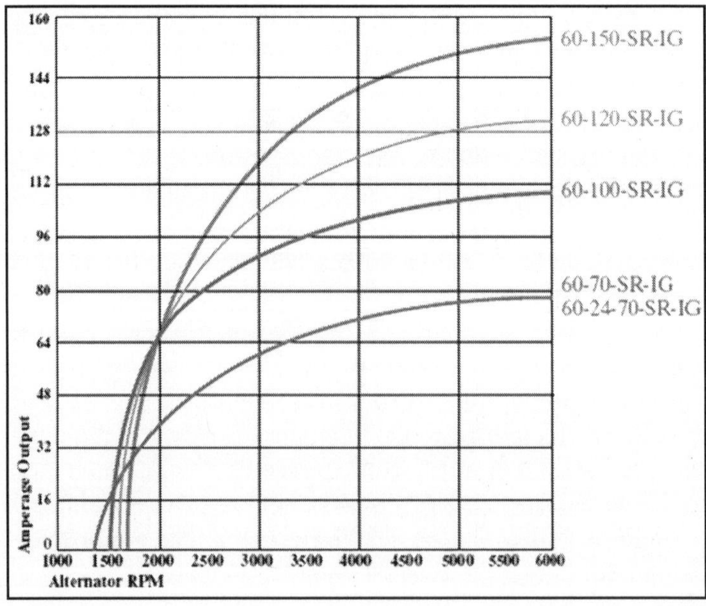

Apart from noting the alternator voltage you should also look to see if you have an isolated ground engine; these are sometimes found on engines with saildrives or on engines with electronic controls. To determine if you have an isolated ground engine configuration just check for continuity between the engine block and your battery negative terminal using a multimeter. If you have continuity then it is not an isolated ground engine and you can use any alternator, otherwise you have to use an special isolated ground alternator.

Sample output curves for alternators at various RPM. Note that the RPM we are talking about is alternator RPM, typically 2.5 times engine RPM. Typically alternators have their full output rated at 6000 RPM but can continue to spin up to 12,000 RPM or more without any additional increase in output.

CHOOSING YOUR ALTERNATOR

Now that you have thoroughly examined what you currently have you are ready to select an alternator. You need to start with an appropriate voltage, case size and mounting foot. Then you have to select an output size based on your electrical demands, battery size, and the pulley type you have. If you have only a single 3/8" wide belt you will be limited to about 70 amps for a 12 volt alternator. If you have a single ½" belt you will be limited to about 110 amps for a 12 volt alternator. Everything more than 110 amps for a 12 volt system will require either double belts or a serpentine belt. The picture below shows you the wide selection of alternators on sale at shop.pkys.com

When considering an alternator size you can apply a simple rule: the alternator output in amps multiplied by 4 should match the size of the battery banks in amp hours. So a 70 amp alternator would work well with a 280 amp hour battery bank. This is a very general rule so don't worry if you bank rating exactly. If your alternator is too big it won't hurt anything but it may not be able to work at its full potential. If your alternator is too small it will just take a bit longer to charge the bank fully.

CHOOSING YOUR REGULATOR

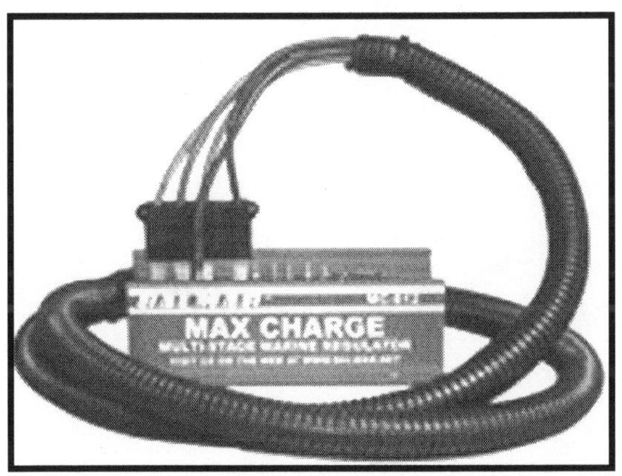

Most high power alternators are designed to be used with an external regulator. An external regulator enhances the alternators performance by allowing the user to make adjustments to compensate for battery type, temperature and other important operational parameters. Some regulators have digital displays that will tell the user what is happening, others use more cryptic indicator lights. Some regulators allow for remote voltage sensing at the batteries for increased accuracy and allow for temperature sensors for alternator and batteries. The alternator temperature sensor prevents the alternator from overheating in the event of sustained high-load use. The battery temperature sensor modifies the output of the alternator depending on battery temperature; a higher voltage for cooler batteries and vice versa. Also worth serious consideration, having a battery temperature sensor in the system can prevent a battery meltdown in the event of a bad cell. The video below shows an ARS regulator in action

INSTALLING THE ALTERNATOR

I normally like to start with the physical installation of the alternator on the engine. It should be a case of just bolting it in, but occasionally there can be small problems. Often you will have to find a longer mounting bolt than the original one. It is customary to use high-tensile steel for these bolts. Occasionally you will have to make some changes to the adjustment arm. You may also need some spacers or washers to ensure the pulleys line up perfectly. In most cases though, if you have carefully selected the correct alternator, it should be a straight fit with few complications. Don't forget to read my blog post on the subject of installing an alternator to get a description of an actual project.

INSTALLING THE REGULATOR

All regulators basically follow the same wiring scheme with only minor differences between them. There is a negative wire (sometimes two) that is usually black. It goes to the alternator negative connection. There is a positive wire, usually red, that goes to the alternator's positive output connection and which should have an appropriate fuse in it at the alternator end (normally 10 amps or less on a 12 volt system). There is a wire that is used to turn the regulator on, usually brown. This can normally go to whatever wire in the old installation was energized whenever the engine was running. And finally there is the field wire, normally blue, that is the output of the regulator and goes to the alternator field connection. This wire provides a variable voltage to the rotating part of the alternator and thus determines how much power it generates.

WIRING THE ALTERNATOR

You will probably need new positive and negative output wires for your alternator because it is unlikely that the original wires were sized appropriately for the new increased output. If you have an isolated ground alternator there will be no electrical connection to the engine block and the negative wire will have to carry the full output of the alternator. Even if you do not have an isolated ground alternator the connection between the alternator and the engine block and is not a good one and is not capable of carrying high loads.

A dedicated negative output wire needs to be installed. In the case of non isolated ground alternators this wire can run to the engine negative connection where it connects to the battery. For isolated ground engines it needs to run to the battery. The positive wire needs to run to the battery too of course, there are various options for this that will be covered in a future article. No matter which way the positive output wire is run it needs circuit protection at both ends in order to comply with ABYC standard E11.10.1.1.2. This circuit protection should be higher than the alternator output to prevent accidental blowing of the fuse, but no more than the current carrying capacity of the wire. Follow the alternator manufacturer's instructions for the output wire size and distance to the batteries.

If your alternator provides a signal to the tachometer then you will have to connect up a wire for that too. This is usually the white wire leading out of the alternator and comes from the stator. It provides a varying frequency signal related to the rotation speed of the engine. You will have to figure out which of the original wires that went to the old alternator to connect this to. You may also have to calibrate the tachometer. This requires a handheld optical tachometer for an accurate reading but a rough and ready approach would be to note what speed the engine idles at and what speed the redline is at no load before you start the installation, then adjust the tachometer accordingly on completion. There are normally a limited number of choices and only one will give a reasonable answer.

FINAL ADJUSTMENTS AND TESTING

When the alternator is physically installed, its output wiring connected and its regulator programmed you are almost ready to start the engine. First make sure the belt is tensioned adequately, then check at the alternator output terminal that you can read battery voltage, alternators won't work if they aren't connected to a battery. Make sure loose wires, clothing and limbs are clear of moving parts and then take your new alternator for a test drive. If you have installed a new belt you may need to retention it after about 30 minutes use. Check belt tension regularly.

ADVANCED PROGRAMMING OPTIONS, TROUBLESHOOTING AND MAINTENANCE

Many regulators offer advanced programming options apart from the basic ones for different types of battery. It is possible to "detune" and alternator that is too powerful by adjusting the software. This is useful if you find that high loads are causing excessive wear on your belts of if you worry that the alternator might take too much power from the engine.

Troubleshooting an externally regulated alternator is easy. First make sure you haven't blown any fuses by ensuring that you can still see battery voltage at the output terminals of the alternator. Then test to see if the alternator is working by momentarily connecting a positive voltage to the field terminal. Be careful not to get in contact with any moving parts. There should be a very noticeable slowing down of the engine as the full load of the alternator is applied.

If the alternator works then test the regulator. If it is the type with a digital display it will tell you what it is doing. Otherwise is it lit up and turned on? If not are you getting voltage on the power wires, the voltage sense wire(if fitted) and the brown wire that turns on the regulator? If all of these are working you should get a variable voltage on the field wire output (normally blue) of the regulator, otherwise the regulator is defective.

12 VOLT ELECTRICAL SYSTEMS

The majority of boats on the water today depend on electricity to some degree. Unfortunately, too many of those boats are operating with inadequate and poorly installed electrical systems. This can result in unnecessary problems that drain your patience and pocketbook and that can cause safety issues. A quick fix may alleviate the problem for the short term, but if the system remains inadequate, problems will continue to plague it.

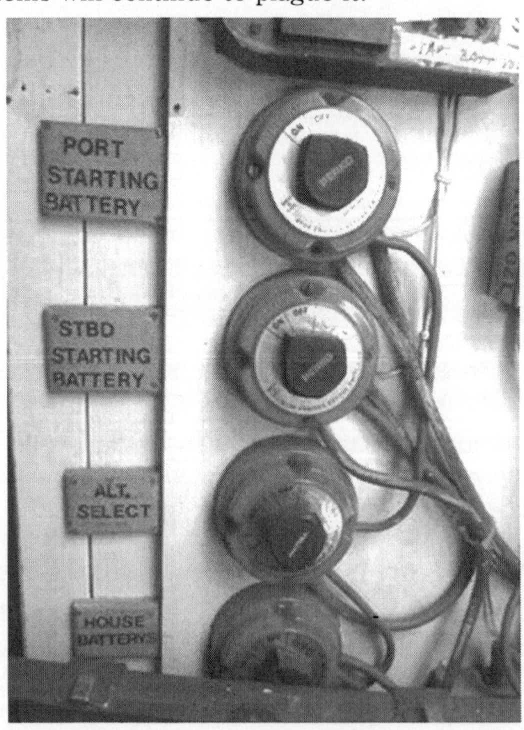

To begin taking the right steps to achieve a good electrical system on your boat you need to:

- Determine your power requirements
- Provide adequate power storage and starting capacity
- Provide the proper charging that you need
- Be sure your wiring is adequate and approved for marine use and connections are sound
- Engage a qualified professional if needed
- Be sure that all safety requirements, such as the relevant ABYC standards, are met

Your boat's batteries are a corner stone of a good electrical system. Power boats that are used simply to run from dock to dock on weekends may need only one (always two are preferable) starting battery which we'll discuss below. These are less expensive and lighter than batteries designed for deep cycling, which is what you'll need if you stop for periods of time to anchor or drift or, far more importantly, if you plan to anchor overnight. Boats that are used for long term anchoring, as in serious cruising, need batteries capable of deeper discharges and of greater storage capacity. Such batteries include wetted lead acid, gel and AGM deep cycle types. However, even if you just plan to run from dock to dock or dock to fishing ground or beach and back, it's important to realize that you may be using electricity during those times that the engine is off and be prepared with the right type and size of batteries.

Batteries need good wiring to deliver the power supply to the gear onboard. That's the subject of "Wiring". Since batteries are essentially fuel cells, charging the batteries is another important element. For additional detail on that subject click on the section entitled "Battery Chargers." There are both DC (direct current as from 12 volt batteries) and AC (alternating current like what you use in your house) systems and issues on many boats, although some boats, particularly smaller ones, only use DC. Here we'll discuss primarily DC. AC options are discussed in some detail in the section "Inverters".

DETERMINING YOUR 12-VOLT POWER REQUIREMENTS

First, calculate your daily (24-hour) average power consumption for all of the electrical loads you place on your system. List all of the appliances and their amp draws. If amps are not listed on the appliance, you can figure amps with the following formula:

$$\text{amps} = \frac{\text{watts}}{\text{volts}}$$

Next, estimate the normal daily usage for each in hours, so that you have a list of appliances and their daily draw in amp-hours (Ah). Now total them all up. But don't cut it close. As battery voltage lowers with us amperage draw will go up. Also, amperage draw increases with wire run. This means that the total of the power consumption numbers on your equipment will probably be less than your actual usage. For further amplification on this see the section on Inverters.

WHAT SIZE BATTERY DO YOU NEED?

Let's assume your daily power consumption totals 100 Ah. A 100-Ah battery won't do it. Why? Because battery capacity is determined in part by the intervals between battery charges, and the discharge level. A 100-Ah battery might in theory meet your daily energy requirements, but would have no reserve. And a battery should never be fully discharged. It must be able to store and deliver the full 100 Ah between charges. Further, with each partial discharge of your battery its capacity will gradually diminish.

Automotive starting batteries are made for starting engines, with the quick release of a big burst of power. They discharge only about five percent, and are intended to be immediately recharged by the alternator. They cannot handle the repeated deep discharges typical of marine use and they may suffer damage from the constant pounding they receive at sea. Conventional wet (lead acid) deep cycle batteries or deep-cycle gel cells or AGM batteries are best at withstanding the deep discharges, rapid high power recharging, and the physical pounding of the marine environment. Look for batteries with the greatest number of life cycles at 50% discharge and do not mix batteries of different construction type (such as, for example, wetted lead acid and AGM) in the same bank - use one or the other. There are also dual purpose batteries claimed to provide relatively high engine starting power for their size but also relatively good deep cycling capability. (Don't count on these for cruising with repeated heavy deep cycling.)

You will possibly double the life of your battery if you don't discharge it below 50% capacity. Consider also, for example, that lead acid batteries recharge rapidly only up to around 70-80% of capacity. If you don't want to spend a lot of time recharging that last 30%, plan on using only about 30% of the battery's full capacity. And it's better to consume even less than that because no battery operates at 100% over its full life. Also, give yourself a little power in reserve. Given all this you, in theory, will need a 400-Ah wetted lead acid battery to meet your 100 Ah daily energy habit. In general, a battery rated at four times your daily usage will be adequate. But, as noted, this is only theoretical. Other factors, such as, for example, voltage drop (and therefore amperage increase) caused by long wiring runs, will affect the usage and size requirements.

BATTERY BANKS

Reserve an adequately-sized, fully charged battery or bank of batteries solely dedicated to starting your engine. Some prefer to use what are commonly referred to as "starting batteries" dedicated to engine starting. These typically have thinner plates but more plates and therefore can deliver more Cold Cranking Amps (CCA) for size and weight. CCA is the measure of how many amps the battery will supply to the starter motor for 30 seconds continuous at 0°F while maintaining its voltage above 7.2 volts. Marine batteries may also be rated in Marine Cranking Amps (MCA) which is similar to CCA except at 32 degrees F. Starting batteries are typically not as tough in construction as batteries designed for deep-cycling and less expensive. These will not serve well if you need them for deep-cycling because such will quickly deplete them, leading to early failure. A deep-cycle battery can be used for engine starting, as long as it provides enough cold-cranking amps (CCA) to easily start your engine.

Use a separate bank of deep-cycle "house" batteries to supply the rest of the boat's electrical needs. Or, you can use two banks of deep-cycle batteries, each with enough cold-cranking amps to start the engine. Parallel the two banks with a dual-purpose battery isolation/selector switch for tough engine starts and then alternate between the two banks for "house" use. Always, however, keep one bank fully charged for starting.

BATTERY CHARGING

If you don't replace what you use, your batteries will eventually go dead, no matter how much battery capacity you have on your boat. The rate at which you can recharge your batteries depends on a number of factors including how much you discharged them, the condition of the battery, the temperature, the alternator's power (in amps) and its output (in volts) as well as an AC battery charger's capacity and features.

Many boats charge their batteries with an engine-driven alternator. If your engine running time is minimal, you want to charge as quickly as possible, without damaging the battery. Battery damage begins when the internal temperature becomes too high, causing it to gas and heat up. If it feels warm to the touch or if you're smelling gas, it's getting too hot. A smart voltage regulator for the alternator should control and taper off the charge to prevent this and can also be set, within limitations, to accommodate your needs. More on this below.

The marine alternator is generally more suitable for use on a boat than the typical automotive alternator sometimes found as original equipment on new gas and diesel engines. Alternators are typically rated in amps; the rating refers to the maximum output in one hour at a certain temperature and rotation speed. Typical automotive alternators can't run continuously at full load in the warm temperatures found around boat engines, or in warm climates. Further they have internal voltage regulators that are set to deliver a rapid charge to a starting battery, not to deliver sustained tapered intelligent deep cycle recharge for house batteries. A suitable marine alternator may be rated at full continuous load at temperatures up to 200°F. Some marine alternators have special grease in the bearings and other special construction to withstand the higher running temperatures and loads as they continue to put out at higher levels during deep cycle recharging. They are also more robust in many other respects.

A good rule of thumb is that you will need at least 120% of the energy you used to restore the battery. And to prolong the life of your alternator, plan on running it at less than full output. Take into account any other power-draining loads you might be adding to the system as you are recharging, such as refrigeration or inverters. If you install an oversized alternator, you can recharge efficiently while at anchor, with the engine at idle and the alternator operating below its rated speed and output. (However, it isn't helpful to the engine to run it for long periods of time under low load. Many prefer to charge batteries while at anchor using an AC generator and AC battery charger. More on this below.) Again, a "smart" regulator which you can set to perform optimally for your parameters is needed.

The speed of your alternator's rotation is a function of engine RPMs and your alternator's pulley wheel size. Once you have determined the maximum alternator output you require, add around 25% or more so you won't have to operate it at full bore to achieve the required results. Now check how many alternator RPMs it takes to reach that output. Then determine the minimum engine RPMs at the engine's power take off wheel at which you will be charging. You need a pulley ratio that gives you maximum required output at your minimum engine speed. If you need 110 amps, and it takes an alternator RPM of 4,000 to generate it, and your engine is running at 1,000 RPMs, then you need a 4:1 pulley. Make sure that if you punch your engine up to 3,000 RPMs, thereby increasing your alternator speed to 12,000, you are not exceeding maximum safe alternator speed. Typically, the manufacturer of a good marine alternator should provide this information for your application and also the alternator pulley wheel that you need.

VOLTAGE REGULATION

A voltage regulator determines the rate and manner which an alternator or AC charger will charge your battery. Normally we think of voltage regulators specifically for alternators, but even AC chargers have internal components that determine rate and manner of charge. In better units these internal components can be field adjusted. If charging isn't done properly your batteries can suffer failures ranging to total depletion to overheating, gassing and essentially self destructing.

The basic alternator voltage regulator maintains voltage at a certain level by matching alternator output with the load and the charge level of the battery. Voltage drops when a load is placed on the power system, or when the battery discharges. The regulator then increases the amperage output of the alternator until the voltage level is restored, and then tapers output to a level that will sustain that voltage.

You should have a regulator that is external, and field adjustable, so that you can adjust the alternator's performance, tailoring the alternator's output to the type and size of battery(s) that you are charging. A "smart" voltage regulator can also be set to charge specific types of batteries in a way that will not only charge them well, but also give them longer life. A good marine adjustable smart voltage regulator should be able to charge at three stages, often called "bulk, absorption and float." Bulk puts a lot of power into the battery rather fast, to bring it up rapidly to a certain point. Then the regulator will switch to absorption mode before the battery gets too hot or gasses or becomes otherwise damaged.

At absorption, charging occurs more slowly to provide more subdued charge commiserate with the state of charge of the battery. Once the voltage is at the desired level the regulator shifts into float which essentially maintains the battery, making adjustment for usage. Some regulators even have heat sensors on the bank(s) so that they can compensate for that factor. Of course all of this must be set for the parameters of your battery and/or bank. Different sizes and different types (such as lead acid, Gel and AGM) of batteries require different types and rates of charging, as well as different voltage levels. Usually instructions with the voltage regulator will help you with this. Your 12 volt system can also be used to supply AC current, like you have in your house, with the addition of a marine approved inverter which will invert the DC current (normally obtained from your battery bank) to AC. Inverters increase the load on the battery and bank size and alternator charging must compensate for this. Of course, the greater the AC draw from the inverter, the greater the inverter's DC draw from the battery. Also, as the inverter is converting at higher amps, its efficiency will diminish, again causing higher rate of battery depletion. This makes it even more important to have a robust marine alternator controlled by an external adjustable voltage regulator. With suitable equipment, you can use AC current while you are running and keep the batteries topped up. If the battery banks are large enough for your consumption and properly charged, you can have quiet times at anchor with AC power available without running the generator all the time.

There are now AC powered battery chargers on the market that can charge and maintain not only different banks but also banks that have different construction techniques, such as wetted lead acid, gel and AGM. These essentially jump from bank to bank in sensing and do what's needed for that bank. Of course, you must set these up appropriately for your system. When you charge your batteries when AC is available (from the dock or from an onboard marine generator) through a battery charger, it should be rated for marine use and also have "smart" charging with similar charge settings and charge characteristics discussed above as to the alternator's voltage regulator. With multiple battery banks many people prefer to have a device which automatically shifts charging output to different banks so that no one bank is overcharged. However many prefer to monitor bank status manually and use a battery selector switch to direct charging current to the appropriate bank. Failure of automatic devices, often hidden away in an engine space, can result in improper charging leading to discharge or overcharge which will "cook" a battery, ruining it, and which could cause explosion or emission of large amounts of corrosive explosive gas.

WIND AND SOLAR CHARGING

Many people also use wind generators and solar cells to keep their 12 volt system charged. Solar cells generally have less output for the money than wind generators and require special mounting racks or cabin or bimini top space. Wind generators can usually put out more current, but only if the wind is up. They make noise, which some find objectionable but others find soothing. Usually people become accustomed to it. Great care must be taken to avoid being hit by a propeller and also to shut it down when the wind gets too high. Better wind generators have a self contained automatic dampening mechanism to prevent over-speed in high winds. One or both systems are often used by those who don't consume much electricity or who want to keep their batteries up while they're off the boat, particularly as when there's no dock power available as when they're at anchor or on a mooring. Even with these systems you must provide overcharge protection. Devices that do this usually come with the wind generator or solar cell.

We can only scratch the surface of this subject, which could easily fill a large book. If you want to be hands on (and that's good) read some of the many books on the subject. In any event, it never hurts to involve a well trained competent professional when dealing with any kind of electricity on a boat. Technology, knowledge and practices change almost daily therefore it is prudent to research for the very latest up to the date information and seek qualified professional assistance when needed.

AUTOMATIC VOLTAGE REGULATOR FOR GENERATOR

The AVR (automatic voltage regulator) is suitable for supplying power to the fields of either brushless or commutator type exciters and will maintain the A.C generator voltage within plus and minus 1% of the preset value with paralleling. When the load varies between no load and rated load and with a frequency changes more than plus minus 4%.Upon application or rejection of rated load,the generator voltage recovers to plus and minus 1% within 0.3 second on most AVR exciter in A.C generator combination.

The AVR is suitable for operation in ambient temperature up to 50 'c and will deliver a power output of 150 watts with a max. current of 3 A.Under some circumstances the A.C generator wave shape changes considerably from no load to rated load. It is due to the type of load,hence the r.m.s average or peak voltage does not stay in proportion.In this situation,i.e: greater than 10% harmonic distortion,the voltage regulation may be outside the limit quoted.

HOW THE AUTO VOLTAGE REGULATOR (AVR) OPERATE?

When rated load is switched on to A.C generator the voltage fall,typically to approximately 70% of its no load value.If the A.C generator voltage is to be maintained near to its no load value,then its excitation must be increased progressively in proportion to the amount of load or more correctly in proportion to the fall in voltage.The AVR perform this task. The voltage to be regulated is derived from the line UR and VR of the A.C generator and is potentially divided by the resistance chain R1,R2 and R3.The signal appearing between terminals T2 of potentiometer RV1 and the junction of C1 and R3.

Rectifier diodes D1 and D2 rectify the signal voltage.An error signal is produced by the difference between the signal voltage and a stabilized reference voltage developed across the zener diodes D9 and D10.Capacitor C1 smoothed the signal.This error signal is applied to the base emitter circuit of VT1.The A.C generator voltage is adjusted by the potentiometer RV1.

ELECTRICAL GENERATION

THE ALTERNATOR

The synchronous machine consists essentially of (a) a field system excited by direct current and (b) an armature. Almost invariably the armature is the stationary member and the field system the rotating member. The induced e.m.f. in the armature winding is a motionally induced e.m.f. and its mode of production identical with that of the D.C. machine. The only difference is that it is the magnetic field which moves whereas the armature conductor is stationary.

As with a D.C. machine the e.m.f. induced in an individual armature coil ia an alternating e.m.f. and consequently by bringing the winding out to fixed terminals, the e.m.f. between these will be alternating also. The complete fixed armature, that is magnetic core and windings, is called the stator, and the rotating field system the rotor. The general constructional features of a salient pole alternator are shown below

As the field system rotates and carries its flux with it, each portion of the stator core will experience reversals of magnetisation, and therefore, as in a direct current machine, the core has to be laminated. For ventilation purposes, a series of radial ventilating ducts are provided. Since the field system rotates, its exciting winding has to be fed by means of two slip rings, but as the excitation voltage is low and the power taken by the field winding small, these present no difficulties.

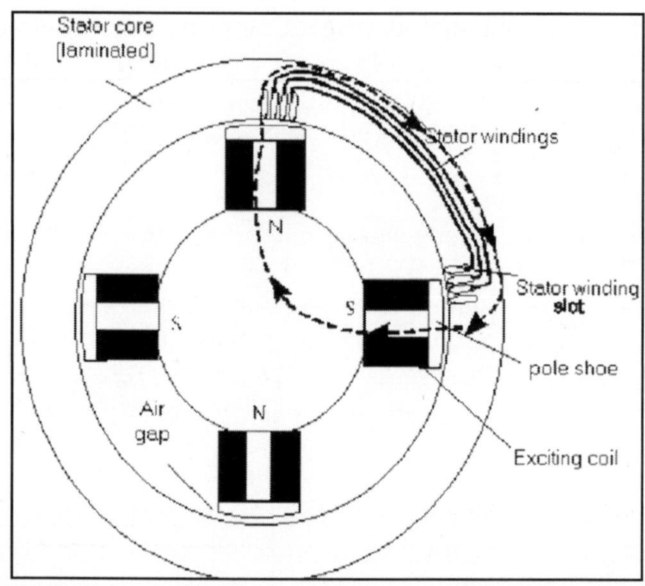

A salient pole has one field coil per pole, very like a D.C. machine. For the very high speeds of turbine driven alternators it is necessary to adopt a cylindrical construction for the rotor and in such a case the field winding has to be housed in a number of slots. A simplified form shows the cross section of a four pole turbo alternator, the dispostion of the rotor field of a turbo alternator rotor may be as high as 40,000 ft per min or 200 metres per sec. The stresses due to centrifugal force are exceedingly high. The rotors are thus made from steel forging, or in some cases from thick steel discs bolted together.

HIGH SPEED ROTOR

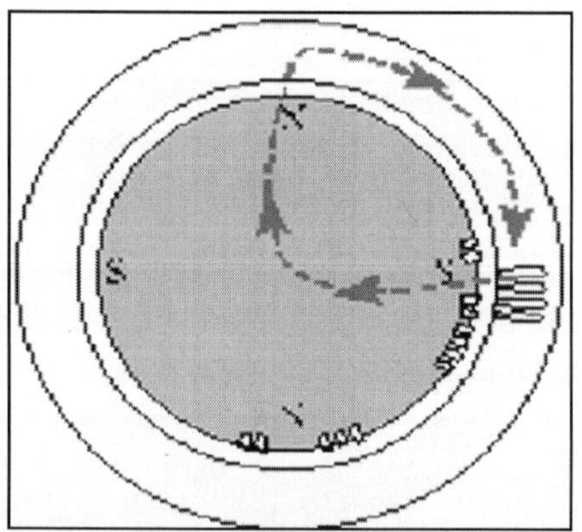

The axial length is normally considerably greater than the diameter. Has the advantage of great strength and stiffness. The exciting current is carried by bar type conductors in the groups of slots shown below. All currents in one group are in the same direction, those on the next group on the opposite direction. Flux produced is distributed over surface approximately according to sine law.

DETAILS OF STATIONARY ARMATURE ALTERNATORS

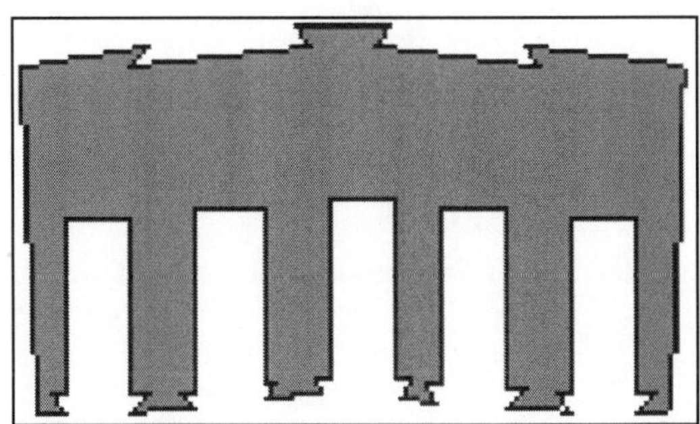

Armature stampings pressed out of sheets of special magnetic iron or steel alloy. In the smaller sizes the stampings are pressed out in compete rings

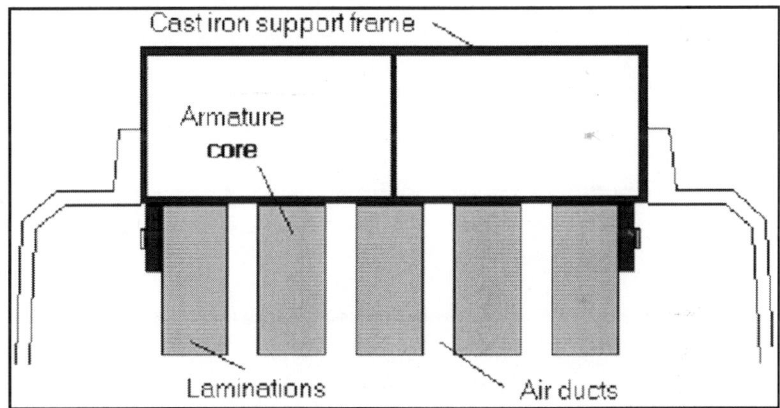

Section through top stator of salient pole machine. The armature core is built up of laminations which are held tightly together by end clamping rings. Spacing strips inserted at intervals leave ducts for cooling air to pass through. The air is driven through by the fan action of the rotor and escapes via the apertures in the cast iron supporting frame

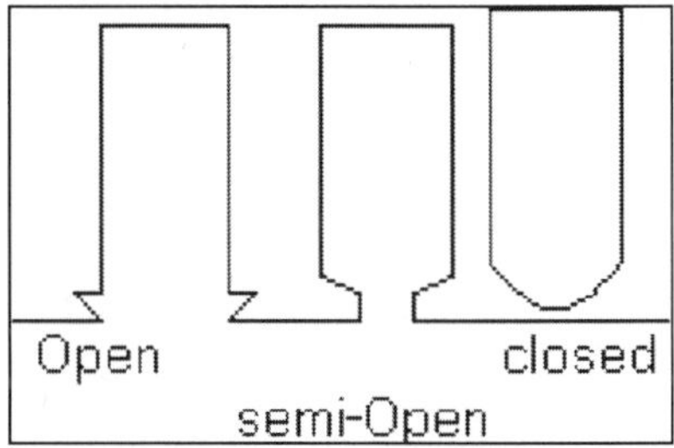

Types of armature slot. The filled slot has round wires but it is common to have rectangular conductors to economise slot space.

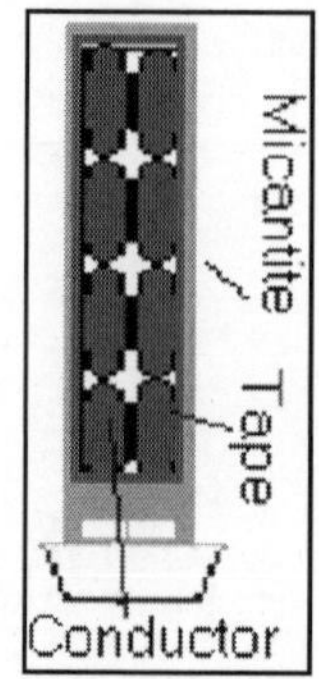

SECTIONAL SIMPLIFIED DIAGRAM OF A TURBO ALTERNATOR

The rotor is turned from a steel forging ans slotted to carry the exciting windings the slots being arranged as shown above. Because of the high running speed, alternators for large outputs have a considerable axial length compared with rotor diameter.

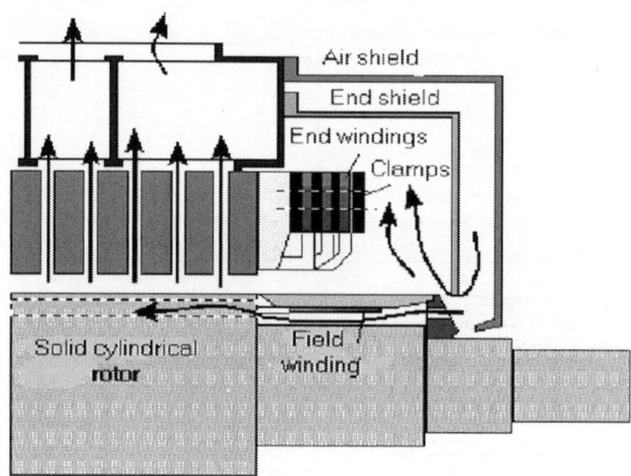

LAYOUTS OF A.C. GENERATORS

CONVENTIONAL EXCITATION SCHEME (ROTARY)

Separately excited D.C. exciter (Out dated)

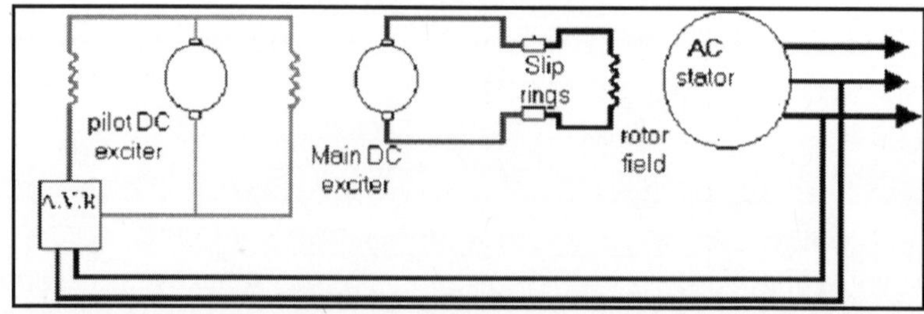

Fig. Brushless excitation scheme using shaft mounted diodes (Rotary).

Indirect self excitation (Error)

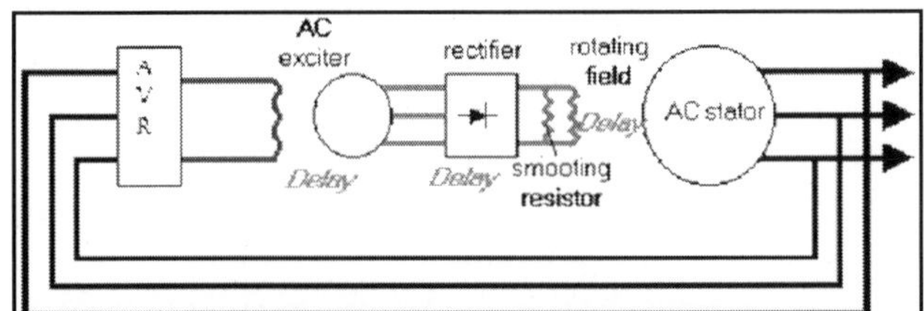

Comparison of the value required to control with a fixed value. When the variable differs from a fixed reference value an 'error' exists and the function of the controlling medium is to restore equilibrium e.g. if the voltage output falls on the brushless rotary excited alternator the a.v.r. controls the exciter field to restore equilibrium.

MODERN COMPOUND SCHEME (STATIC)

Direct self excited (Functional)

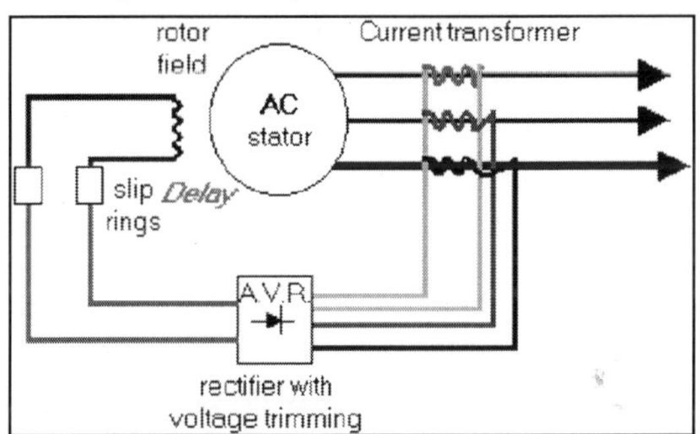

Control of the voltage to a set value is achieved by the inherent characteristics of the machine. A compound wound d.c. generator with a level compound characteristic has additional current in the series field under load conditions. In the self excited compound alternator there is a constant amount of excitation required for no load condition. Additional excitation due to more current form the current transformers is obtained in response to extra external demands

RECOVERY GRAPHS FOR FUNCTIONAL AND 'ERROR' LAYOUTS

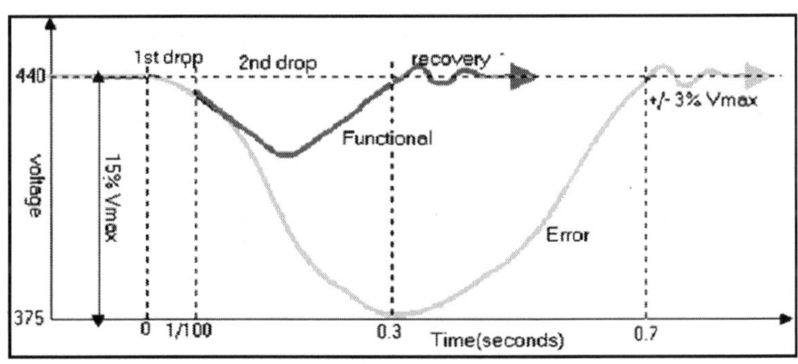

SHAFT DRIVEN GENERATING SYSTEM

Methods of drive

1. Belt or chain driven
2. Direct coupling engaging the propeller shaft
3. Power taken from the main gearbox.
4. Power taken from the free end of the engine

With D.C. auxiliaries power can be taken by either a chain or belt drive from the propulsion system with an A.V.R. maintaining constant voltage.

For A.C. systems methods used include the use of a D.C. generator with an D.C./A.C. converter, or direct A.C. generation. With the latter either a constant speed drive is required or a frequency converter. With either method the revolutions at which the shaft alternator can be used is limited. In this way direct drive systems will generally be fitted in conjunction with a C.P. system which maintains constant engine speed under full away conditions.

Advantages

- Saving on fuel costs, allows efficient use of heavy rather than gas oil
- Reduced maintenance costs
- Capital saving on reduced number of auxiliary sets
- Reduced space and weight
- Reduction in noise

Disadvantages

- Power available for propulsion reduced
- Capitol cost of plant
- Auxiliaries required for manoeuvring, although some medium speed plants are capable of manoeuvring with shaft alternators and C.P. system
- Complicated constant speed or frequency gear required with slow speed engines

AIR GAP

If the air gap around a rotor is not uniform the motor may not start in certain positions. Because the rotor is not centred, probably due to worn bearings, there is an out of balance magnetic pull. Radial play in between the shaft and the housing should be detected by hand and bearing wear detected by feeler gauge between the rotor and the stator, or armature and field poles may be measured at three or four fairly equidistant points around the machine. If possible one measurement should be made at the bottom of the machine and another in line with the drive. Compare with previous records to check wear. At minimum air gap. Clearance of the bearings should be renewed to avoid the possibility of the rotor rubbing on the stator.

On small machines two feelers on opposite sides of the rotor should be used to avoid error caused by rotor movement from normal position when only one feeler gauge is used. In synchronous motors and D.C. motors sparking may occur if the radial air gaps between the armature and the field poles are unequal. If necessary renew bearings or add or remove soft iron shims from under the pole shoes. Unequal field strength has a similar effect of sparking at the brushes. This might be due to short circuit or earth fault on the field coils, or a short circuit on the shunt and field coils.

An increase of air gap gives an increase in 'reluctance'. In a salient pole A.C. generator this fact may be used to produce a sinusoidal flux density curve by gradually increasing the length if the air gap towards the pole tips.

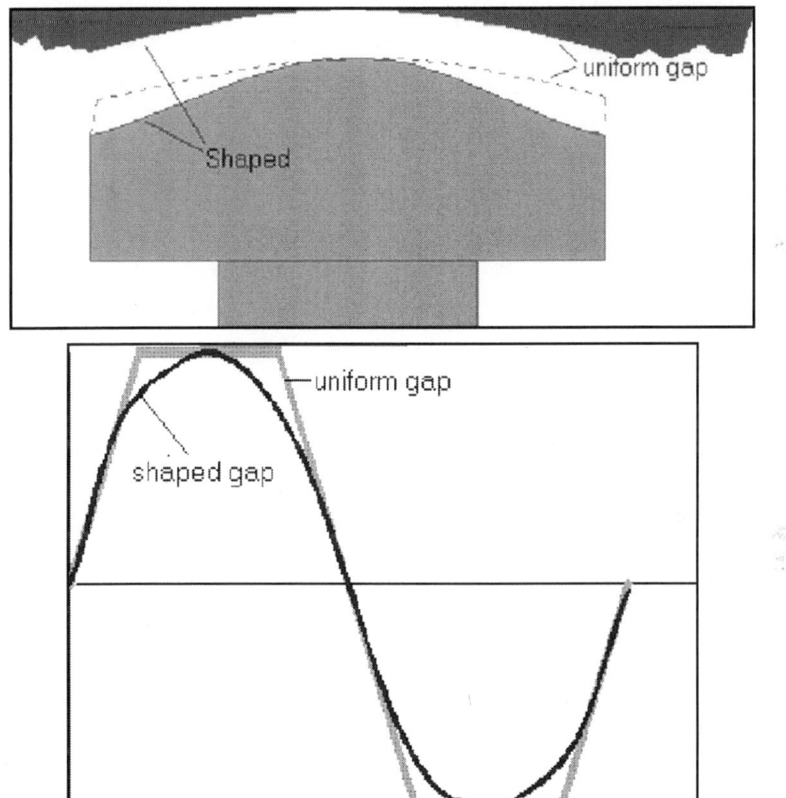

In the induction motor the air gap should be as small as possible if the motor is to act with a high power factor. An increase in air gap increases the reactance of the motor and lowers its power factor. Small motors are accurately machined and centring of the rotor is very important so ball or roller bearings are fitted.

Air gap Motor size
0.25mm 1kW
0.75mm 10kW
2.0mm 100kW

PARALLEL OPERATION OF GENERATORS

D.C. GENERATORS

For compound wound D.C. generators it is usually sufficient to ensure that the voltages of the incoming generator is the same as the bus bar voltage. The equalising connection joining the junctions between the armatures and their series fields is incorporated in the circuit breaker in such a way that the equalising connection is automatically closed before and opens after, the main contacts. By adjustment of the shunt field regulator the load sharing may be controlled

A.C. ALTERNATORS

To parallel alternators the following conditions are required;

Same voltage-checked with the voltmeter

Same frequency-checked with the frequency meter and synchroscope

Same phase angle-checked with synchroscope

Same phase rotation-checked with rotation meter. Only important when connecting shore supply, or after maintenance on switchgear or alternator.

LOAD SHARING OF ALTERNATORS IN PARALLEL

Alternators in parallel must always run at the same speed. After a machine has been paralleled and is required to take up its share of the load, this will not be achieved by adjusting the field excitation current. Although the increase in e.m.f. will cause a current to flow in the busbars, and this will show on the machines ammeters, this is a reactive current that lags the e.m.f. by 90o and produces a reactive (kVAr) but not kW. Its only effect is to alter the operating power factor of the alternator.

More power may be obtained at the bus bars from the incoming alternator only by supplying more power to its prime mover. This increase of steam or fuel supply is achieved by altering the governor setting either electrically or manually. After adjusting the governor the incoming machine takes up its desired amount of the kW loading and this is recorded on the machines watt meter. However, if the kW loading is shared equally between two machines it may be found that the Load Current of the incoming machine is more or less than the other machine. This is fue to the incoming machine having a different power factor. This may be corrected by adjusting the excitation of the incoming alternator.

Thus after paralleling an alternator;

Adjust prime mover governor until kW loading is correct

Adjust field excitation current until current sharing is correct.

If the alternators have similar load characteristics, once adjusted, the load will continue to be shared. If the load characteristics of alternators vary, the kW loading and load current sharing may require readjusting under different load conditions.

VARYING THE DRIVING TORQUE

If the driving torque of 'B' is reduced (less fuel supplied) the rotor falls back by an angle say p.f.(b) giving a resultant e.m.f. of Ez in the closed circuit. The e.m.f. Ez circulates a current I which lags behind Ez by angle p.f.(a). This circulating current Iis more or less in phase with Ea and in opposition to Eb. This means that A is generating power to motor B and this will compensate for any loss of power in the prime mover of B.

Once the power increase in A equals the power loss of B balance is restored and A and B continue to run in synchronism. Therefore the power is shared by adjusting the torque (fuel input.) Slight loss of power in B-is taken up by an increase in power from A. The terminal voltage will not vary and the speed and frequency will stay the same or drop only very slightly. Large loss of power in B-with a large circulating current from A to B the alternator A will try to drive B as a synchronous motor. The amount of full load power required to drive an alternator as a motor is only 2 to 3% for a turbine and 10 to 12% for diesel engine. As the circulating current flows from A to B the reverse power trip on B will operate after about 3 to 5 seconds. All the load now falls on A which will probably cause the overload trip to operate and 'black out' .

VARYING EXCITATION

Consider A and B are exerting the torque required by its alternator and the generated e.m.f. Ea and Eb are equal. There is no circulating current. By reducing Rb the excitation current in the field of B can be increased and Eb will increase. Ez is the resultant difference (Eb - Ea) which will give a circulating current I through the synchronous impedances of the two alternators. As the machines are similar the impedance drop in each will be 1/2Ez so the terminal voltage

$$V1 = Eb - H\ Ez = Ea + H\ Ez$$

Therefore increasing the excitation current will increase the terminal voltage As p.f.(a) is almost 90o the Power circulating from B to A is very small Ez I Cos [p.f.(a)] approx equals Zero (Cos 90o = Zero)

EFFECT OF REDUCING EXCITATION

By increasing Rb the reduction of the field excitation current of B will reduce the terminal voltage

Ea>Eb terminal Voltage V = Ea - H Ez = Eb +H Ez

The circulating current I from A to B will have a large 'Wattless' component. Machine A now has more of the lagging reactive current and its power factor is reduced. Too large a reduction in excitation current in B with subsequent increase in load current in A could cause the current overload trip of A to operate. This could be followed by the low voltage or the overload trip of B operating causing a black out.

VOLTAGE REGULATION

The graph demonstrates that excitation must be increased (generally) with increasing load to maintain terminal voltage

The worse the power factor the worse the terminal voltage change during load change. Voltage regulation = DV when load removed/ Full load terminal voltage

At 1.0 p.f. = AC/ OA

At 0.8 p.f = AD/ OA

Therefore lower p.f. = greater voltage regulation

LIMITING VOLTAGE DIP AND RESPONSE TIME UNDER IMPACT LOADING

The effect of a large load suddenly switched on to a small power installation such as a ships plant will be an instantaneous dip in the generator voltage. This effect, due to the transient reactance on starting, cannot be obviated either in a self regulated machine, or in a conventional generator with A.V.R.

The sluggish response of the excitation systems limits the speed of voltage recovery. In a self excited generator the dip is less and the recovery time greatly improved. (say 0.3s against 0.7s) In order to maintain constant voltage, under varying conditions, excitation must be varied.

ELECTRICAL DISTRIBUTION

AC SWITCHBOARDS

If voltages exceed 250 volts d.c. or 55 volts A.C. then the switchboard must be dead front (no exposed live parts at the front) of the metal clad type.

BUS BARS

High conductivity copper rated to withstand the thermal and electromagnetic forces which would arise in the event of a short circuit at the bus bars with all the generators in parallel. The bus bars will withstand these conditions for the length of time it takes for the alternator circuit breakers to trip or back up fuse to blow.

Certain instruments and controls require a feed direct from the bus bars. Any connection between the bus bars and protecting fuses must be capable of withstanding maximum fault level. Standard practice is to provide a three phase set of fuses, known as 'Back Up' fuses, as near to the bus bars as possible. Connections are then led to the racks of the many instruments fuses fitted.

CIRCUIT BREAKERS

Must be capable of making and breaking under normal conditions and also abnormal conditions such as a short circuit. As the circuit breaker must be able to withstand closing onto a fault conditions without sustaining damage, it is of heavy construction. Fitted with an over current release and overloads with time lags, a circuit breaker can be used as follows;

- To control the output of a generator
- As a direct on line starter
- Control outgoing feeder circuits

On modern switchboards 'draw out' circuit breakers may be fitted. In the open position the whole circuit breaker can be wound clear of the bus bars, thus full inspection and maintenance can be achieved without the necessity of de-energising the bus bars so providing a separate isolating switch. The 'plug in' contacts joining the circuit breaker to the bus bars are not capable of taking the breaking load and it is essential that the circuit breaker is in the open position before any attempt is made to withdraw it. A mechanical interlock is fitted arranged to trip the circuit breaker before the winding handle can be inserted,

The breaker also has a mid position, in this position the control circuits are still connected with the bus bar connection isolated. The electrical operation of the breaker can then be tested. Circuit breakers are normally fitted with under voltage protection and tripping is accomplished by shorting or open circuiting the no-volt coil which releases the latching in mechanism. The no-volt coil may also be open circuited by a reverse power relay and an overload trip fitted with a time delay

INSTRUMENTS

The following instruments are the minimum that must be fitted:

- Bus bar voltmeter and frequency meter
- Volt meter and frequency meter, with selector switch to measure incoming machine conditions
- Ammeter with phase selector switch for each alternator
- Watt meter for each alternator
- Synchroscope and if check synchroscope not fitted lamps
- Earth leakage indicator

Additional instruments that may be fitted

- Watt hour meter
- Power factor meter
- Alternator excitation ammeter
- Alternator excitation volt meter
- kVAr meter
- Share connection supply meter
- Emergency batteries on discharge meter

When a check synchroniser is fitted it is there to prevent connecting an incoming machine to the bus bars whilst out of phase, it is not there as aid to synchronising. In an emergency the 'in synch' light may be used to indicate when the breaker may be closed.

When an incoming machine is selected, its no-volt coil and circuit breaker contactor relay coil are connected in series with contacts on the check synchroniser.

These contacts must be closed, that is the machine in phase with the bus bars, before the breaker contactor relay may be energised. If starting from a dead ship the check synchroniser must be switched to off before the first generator is put on the board.

PROTECTION

- Overload protection-fitted to circuit breakers
- Reverse power-When motive power is removed an alternator will try to become a synchronous motor and draw current from the circuit. A reverse power relay will operate after about 2 seconds and about 2-3% reverse power for turbines, 10-12% reverse power for diesels. The time delay prevents tripping during paralleling and taking the alternator off the board.
- Preference trip-automatically, and sometimes sequentially, sheds load from board to maintain supply to essential services during periods of overload.
- Fuses-Usually of the HRC type
- Discrimination-The protective device closest to the fault should operate and protect other services
- Group starter board-Large demand sections may be separated from the main switchboard by fuses and circuit breakers.

AUTOMATIC VOLTAGE REGULATORS

Shall be supplied separately from all other instrument circuits. Protection should be by fuses mounted as close to the supply connections as possible.

Shore supply connections

- Where arrangements are made for the supply of electricity from a source on shore or other location a suitable connection box has to be installed in a position in the ship suitable for the convenient reception of flexible cables, it should contain a circuit breaker or isolating switch, fuses, and terminals of adequate size to receive the cable ends.
- For three phase shore supplies with earthed neutral terminals are to be provided for connecting hull to shore earth
- An indicator for shore side connection energised is to be provided.
- A means for checking polarity or phase rotation is to be provided

- At the connection box a notice indicating ships requirements with respect to supply as well as connection procedure.
- Alternative arrangements may be submitted for consideration.

CIRCUIT BREAKERS

When selecting a circuit breaker for a particular application the principal factors to consider are; system voltage, rated load current, and fault level at the point of installation

VOLTAGE RATING

At medium voltages the phase to neutral voltage may be 250v but the potential difference between two phases with the neutral insulated would be 440v. At these voltages no difficulties should arise in selecting the circuit breaker equipment. However, on a 3.3kV insulated neutral system the phase to neutral voltage is 3.3kV/ 3 = 1.9kV. If an earth fault develops on one phase the potential of the other two phases to earth is 3.3kV. To ensure the insulation is not subject to excessive stress a circuit breaker designed for a normal system voltage of 6.6kV may be fitted. Also on insulated neutral systems high over voltages may be caused by arcing faults. Medium voltage systems switch gear insulation should be able to withstand such voltages, but 3.3kV and above, the margin of safety is reduced. When a high voltage system is installed both the voltage rating of the circuit breaker and the method of earthing must be considered.

CURRENT RATING

Consider three factors;

Maximum permissible temperature of circuit breaker copperwork and contacts temperature due to LOAD CURRENT

Ambient temperature

In industrial use the ambient temperature considered is usually 35oC. If uses in a marine environment temperature of 40oC (Restricted areas) and 45oC (unrestricted areas) are used, therefore the circuit breaker rating may be 'free air' value and this does not consider the degree of ventillation, the number and position of the circuit breakers or the layout of the bus bars. The final switchboard arrangement could be only 80 to 90% of the free air rating

FAULT RATING

Breakers should be rated to accept a breaking current of about 10 times the full load current. The breaker should also be able to make against a fault condition where the making current may be 25 times the full load current when the contact first make. Circuit breakers must remain closed for a short time when a fault occurs in order to allow other devices which are nearer to the fault to trip first. The breaker should be capable of carrying its breaking current for a specified time of usually about one second.

ARC SUPPRESSION

The blow out coils, which are connected in series with the circuit breaker contacts, form an electro-magnetic field which reacts with the arc to give a deflecting force which tends to bloe the arc outwards. The increase in effective length of the arc causes it to extinguish more quickly. The blow out coils are protected form the arc by arc resistant material which may be in the form of an air shute

Hot ionised gases around the arc and contacts are displaced by cold air forming eddy current air flow. This helps to increase resistance between contacts.

CONTACTS

Attention should be paid to all contacts likely to deteriorate due to wear, burning, inadequate pressure, the formation of a high resistance film or becoming welded together. Faulty contacts are often indicated by overheating when loaded. Different contact materials may need different treatment. Copper is widely used but is liable to develop a high resistance film, and copper contacts may become welded together if the contact pressure is low and the contents have to carry a high current. Copper is commonly used for contacts which have a wiping action when closing and opening., this action removing the film. Copper contacts are used on knife switches, laminated (brush) contacts of regulators and other controllers, drum contacts, etc.

Carbon and metallized carbon contacts are unsuitable for carrying high currents for long periods but, as they do not weld together, they are used for arcing contacts on some control gear. Pure silver and silver ally contacts tends to blacken in service but the oxide film has a low resistance. Copper- tungsten (sintered compound), grey I colour, is used in contact facing. This material has a high surface resistance which resists heavy arcing and does not weld. Silver tungsten (sintered) has similar properties to copper tungsten but has a lower contact resistance and is less liable to overheat on continuous load..

SERVICING CONTACTS

Copper contacts should be filed up if necessary to restore the profile required to ensure correct wiping action. Copper contacts which have become burnt or pitted or otherwise damaged, may be carefully dressed with a file. Emery cloth should not be used.

Some contacts are provided with pressure adjustment, so if the contact pressure is reduced by dressing it should be readjusted. Using a spring balance pulled in a direction normal to the contact surface a reading should be taken when a piece of paper placed between the contacts is released. Inadequate spring pressure may also be due to the pressure springs becoming weak due to fatigue or overheating.

Carbon contacts should receive the same attention as copper contacts except that they should not need lubrication. Silver, Silver alloy and copper tungsten contacts do not require cleaning. As there is no need to remove surface film from pure silver contacts they may be used for light butt contacts.

Where some contacts have the appearance of pitting on both faces this is sometimes referred to as being 'burnt in'. Some manufacturers recommend that the contacts, unless there is loss of material, are not dressed as this may destroy the contact area.

AVR'S

R1-Sets volts value

R2-Trimming resistor (Power factor correction)

R3-Trimmer

Carbon pile-Control resistance for AVR

Operating coil-Along with carbon pile form the controlling elements

CCT and PT-Are the detecting elements, the CCT acts as a feed forward device indicating future voltage changes by detecting variation in current flow Stabilising element-Is the capacitor across the Exciter (may be replaced by a resistor) The A.C. voltage is applied to the operating coil through a full wave rectifier. This A.C. voltage supply induced in the potential transformer and the circulating current transformer may vary under varying load conditions such as direct on line starting of relatively large motors. The capacitor connected across the coil smoothes the D.C. output from the rectifier.

If the A.C. applied voltage falls, the field of the solenoid weakens, and the resistance of the carbon pile decreases. With less exciter circuit resistance the current in the exciter field increases thus increasing the output voltage of the A.C. generator. The automatic voltage regulator voltage output may be adjusted with the hand regulator R1 in the exciter field. Before synchronising the alternator the open circuit voltage is adjusted with the hand regulator R1.

After synchronising, and after the kW loading has been adjusted on the prime mover governor, the field excitation under steady load conditions may be adjusted using the Trimming resistor R2. Using the trimming resistor the power factor of the incoming machine will be equalised with the machines already in use. If the load power factor now changes then the terminal voltage will regulate badly, e.g. a rise from 0.8 to Unity Power factor will cause a rise in terminal voltage of about 20 %. So a small Voltage Trimmer R3 is provided across each current transformer to adjust terminal voltage when there is a change in overall power factor

MODERN A.V.R. (ZENER BRIDGE)

Voltage across the Zener diodes remains almost constant independent of current variations. Smoothed D.C. output is applied to the voltage reference bridge. This bridge is balanced at the correct generator voltage output with no potential difference between 'A' and 'B'. If the generator voltage fails, current through the bridge arms falls and current flows from 'A' to 'B' through the amplifier.

If the generator voltage falls, current through the bridge arms falls and current flows from 'B' to 'A' through the amplifier. If the generator voltage rises, Current through the bridge arms rises with current flow from 'A' to 'B' through the amplifier. The signal from the amplifier will automatically vary the field excitation current, usually through a silicon controlled rectifier (Thyristor) control element.

The Silicon Controlled Rectifier (Thyristor) is a four layer, three terminal, solid state device with the ability to block the flow of current, even when forward biased, until the gate signal is applied. This gate signal could come from a Zener diode Voltage reference bridge. The gate signal will switch on the forward biased S.C.R. and current flows through the exciter field. When reverse biased the S.C.R. will again block current flow. Due to inductance of the field winding the S.C.R. would continue to pass current for a part of the negative cycle. By fitting a 'free wheeling' diode the current though the Thyristor falls quickly at the end of the positive cycle. In some circuits the excitation current is designed to be excess of requirements, so that the gate signal reduces flow

INSULATED NEUTRAL SYSTEM

Advantages

- This system avoids the risk of loss of essential services e.g. steering gear
- If the neutral was earthed and a short circuit on one phase causes the fuse in that phase to blow the system would now be singled phasing and may burn out motors
- In an insulated neutral, one earth fault does not interrupt the supply but an earth leakage detection system will give warning.
- Low earth fault currents in insulated systems gives a much less fire risk.

Disadvantages

- On the insulated system the voltage to earth is 1.73 Vph e.g. 440v vs 250v
- Tracing an earth fault is more difficult because although selective tripping may trace the earthed circuit, the actual position on the circuits may still be difficult to locate. Resonant or intermittent faults in say a contactor solenoid or a transformer with an insulated neutral can cause voltages to be magnified to say 4 times the normal voltage to earth (250v $\times$ 4 = 1000v)

Note: electrical shock is not reduced by using a non-earthed neutral as large voltages are involved. Both systems are equally dangerous

EARTHED NEUTRAL SYSTEM

When an earthed neutral system of generation is used earthing is to be through a resistor. The resistor is to be such that it limits the earth fault current to a value not greater than the full load current of the largest generator on the switchboard section and not less than three times the minimum current required to operate any device against

SYNCHROSCOPES

The armature of the synchroscope carries two windings at right angles to each other and is capable of rotation between field poles F F1 R is a non inductive resistance and XL is a highly inductive resistance both connected to one phase of the bus bars. This produces a field which rotates relative to the armature at the bus bar frequency. When the incoming machine is connected to the coils of the field poles a pulsating field is produced at the same frequency as the incoming machine. If the two fields are not at the same frequency then the armature will rotate at a speed equal to the difference

In the modern rotary synchroscope there are no slip rings. The rotor has two soft iron pole pieces and with its shaft carrying the pointer it is magnetised by coil R from the bus bars. With this coil is fixed adjacent to the shaft, therefore, there are no moving coils, contacts or control springs.

THREE PHASE SYNCHRONISING

Synchroscope with Two Lamps (Lamps Dark)

The secondary windings of transformer T1 supplies field coil F of the synchroscope. The secondary windings of T2 supplies the rotating coils R of the synchroscope.

If the incoming machine is in antiphase with the bus bar the voltage difference between the output of the secondary of T1 and T2 is double the normal voltage giving normal volt drop across each lamp. When in phase there is no voltage difference between the outputs of T1 and T2 and therefore lamps are dark when synchronised.

REVERSE POWER TRIPPING

A.C

A non magnetic metal disc can rotate in a magnetic field between two electro magnets. The disc is restrained by a coil spring. The flux produces a torque on the side which rotates the trip lever away from the trip contacts. In reverse power conditions the flux from the voltage coil and current coil interact to rotate the disc in the reverse direction. The amount of torque/current (and hence power) is set on the current coil tapping. A permanent magnet is provided on the disc to provide damping. A 3 to 5 second delay is incorporated into the trip circuit to allow for transients when paralleling.

D.C.

1. Low voltage coil
2. Overcurrent trip
3. Reverse current trip
4. Reverse current trip

2a and 2b are fitted in case of circulating current via the equalising connection. Under normal running, fields of '3' and '4' act together to hold the trip contact down. With reverse current fields are in opposition and a spring pushes the plunger against a trip bar to open the reverse current trip relay.

TIME DELAY DEVICES ASSOCIATED WITH SAFETY CIRCUITS

Dash Pots-(Magnetic Time Lag)

The usual form of time/delay is an oil dash pot having an inverse time/current characteristic, relies for its operation upon the retarding action of a plunger immersed in a reservoir of oil, together with the magnetic force generated by a flow of current through a solenoid. The plunger is attached to an iron core which is partially enclose in the solenoid. When the solenoid is sufficiently energised the iron core will attract it, but the action is retarded due to the oil, in this way a time lag is introduced

Characteristic of Oil Dash Pot with Inverse Time Delay

It is important to note that as the viscosity of the oil varies with temperature, so will the operating times of oil dash pots vary. Makers will supply dash pot oil suitable for the circuit breakers and relays. The recommended oils are selected on the basis of least variation in viscosity over the working range coupled with viscosity's which will give the time delay marked on the calibration plate. These special oils should invariably be used, the time delays are usually calibrated at 15oC, unless otherwise stated and are only correct at this temperature .

Over current devices fitted with oil dash pot time lags do not operate at the current marked on the calibration scale but at a current 25% greater with the appropriate time delay. The current marked on the scale is the value at which they would operate without time delay. Some makers supply an instruction plate indicating the exact current at which the relay will operate with a given setting.

Thermal Device

Depends for its action upon the heating effect of an electric current flowing through, either a bi-metallic strip, or a heating coil placed near a strip. The thermal characteristics of the two dissimilar metals is such that when sufficient heat is generated there is a movement of the strip in one direction until the relay contacts are opened.

Induction Relay

Similar to the action of the watt-hour meter, consisting of a metal disc pivoted so that it is free to rotate between two poles of two electro-magnets. The disc spindle carries a contact which is arranged to bridge two contacts when the disc has rotated through an adjustable angle. A spring returns the disc to the reset position, and as, during the deflecting period, the torque exerted by the spring increases, this is compensated for by the provision of graduated slots in the discs periphery. The necessary damping of the movement is provided for by incorporating a permanent magnet through which the disc has to rotate. The upper electro-magnet contains two windings, one, the primary normally is connected to a current transformer to a winding on the lower electro-magnet. Because of the graduated slots, the inertia of the moving system prevents the disc form rotating under normal running conditions, but when overcurrent commence to flow through the external circuit, the torque generated by the interaction of the upper and lower electro magnets is sufficient to cause the disc to rotate, until either the current falls to a safe level, or the relay is operated

THERMAL INSPECTION OF SWITCHBOARDS

In service inspection of switchboard, individual starter and distribution panels can be carried out with the use of infra red temperature measurement or Thermal imaging systems. They require no direct contact with the electrical components being measured This form of inspection is used to locate areas of increased temperature either associated with local overload or with increased resistance.

INFRARED TEMPERATURE MEASUREMENTS

This takes the form of a relatively inexpensive, readily available temperature reading instrument often with laser guidance. The taking of readings is very simple although care must be taken in the interpretation of results. Due to their low cost these are normally found on most vessels as standard item

THERMAL IMAGING

These tend to be expensive specialist instruments although they may be used by untrained personnel and results are generally simple to interpret. Due to their high cost it is normal to have a single unit available for several vessels delivered as required. The unit is used rather like a video camera and the results viewed in realtime as an image displayed on the rear of the instrument

A typical unit is that seen in common use with fire brigades who use it in search and rescues

DISCRIMATION AND FUSES

Discrimation

A circuit fed from a distribution board may be fed through three or even four fuses or circuit breakers e.g. a heating circuit may be connected to a 15amp fuse in a fuse box fed from a section box in turn from a 500A circuit breaker on the main board. Discrimination occurs when the fuses nearest to the fault operates leaving all the other fuses or protective devices intact. Discrimination may be required between fuse and fuse, or between fuse and overcurrent device such as a circuit breaker.

Fuses

A fuse is a protective device which is there to prevent overloading. If too heavy a fuse or if the fuse is overridden then there is a possibility of overheating, deterioration of insulation and failure. Materials used are; Tin, Lead, or silver having low melting points. Use of copper or iron is dangerous, though tinned copper may be used. Unlike some other forms of circuit protection devices (oil switches for instance), which are suitable for a.c. only, solid filled cartridge fuses have an approximately equal breaking capacity for D.C. and A.C. and the action of the fuse does not depend on breaking circuit at the zero point on the current cyclic wave

Requirements of a fuse:

- Have a high cut off speed on short circuit
- Operate on a sustained overload
- Make a complete break after operation with high insulation resistance between terminals
- Have a maximum temperature rise on continuous full load rating such that it will not be injurious to rudder cables connected to its terminals
- Be mechanically robust and capable of withstanding the large magnetic and gaseous forces exerted during operation under short circuit conditions

A fuse must not:

- Operate on continuous full load
- Operate on momentary overload
- Deteriorate under continuous operation at full load
- Burst or emit flame or otherwise damage the fuse carrier and base when it operates

REGULATIONS

I.E.E. and classification society rules now specify high breaking capacity (high category) fuses on main switch boards where the total normal generator capacity exceeds 400kW at 200v, this is for short circuit or low resistance protection of the very high currents that can be generated in these conditions.

In addition;

- Fuses on shipboard must be made to approved standards.
- Breaking capacity not less than prospective short circuit current at point of installation
- Fuses over 300A are not used for overload protection but may be used for short circuit protection
- Cartridge fuses only on tankers

To control the extent of heavy fault currents on large installations the protective device must have a very high speed performance or High Rupture Capacity (H.R.C.) H.R.C. fuses will operate quickly before the short circuit current exceeds 3 times the full load current.

CARTRIDGE FUSES

Are capable of handling large short circuits. Because of standardisation of manufacture they have very consistant time/current fusing characteristics making them accurate, dependable and non-deteriorating in service. Suitable filling powders such as silicon sand are used in cartridge fuses having the property of quenching the arc of the fused element.

ENCLOSED FUSES

The element usually made of silver is much smaller than the tinned copper used in semi-enclosed fuses so that the amount of vaporised metal is less and this contributes to a better performance. The enclosed casing and use of silver ensures no degradation due to oxidation. After the silver element has fused the indicator wire will heat up sufficiently to ignite the indicator powder and the fuse will be shown to be blown.

Except in the lowest ratings there are two or more elements in parallel which increase the contact area in contact with the filler, and this increases the breaking capacity. The ends of the element are reinforced by larger wires to reduce resistance and therefore heat losses. The indicator type should in the construction below consists of an indicator wire which ignites an explosive powder which chars the indicator paper. On other designs the indicator wire releases a spring and pop up indicator

SEMI-ENCLOSED FUSES

Tinned copper fuse wire exposed to the atmosphere tends to deteriorate and will vary in performance after long periods in service. Also there is a temptation to increase the gauge of the wire, or the number of wires after a fuse has blown. However, rewireable fuses are cheap, easily replaceable, blown fuses are easily detected and within reason if the circuit is uprated slightly no new fuse holders are required.

Tin-fast heating and failure (expensive)

Copper-Slow heating and failure (cheaper)

On overload the tin will fail rapidly increasing the current through the core speeding up its failure.

RATING

Is that current the fuse will carry continuously e.g. for a circuit rated at 30 amp, a 30 amp fuse will be appropriate. Fuses and circuit breakers on switchboards and distribution boards are intended primarily for the protection of the cables and not the apparatus. Overload protection of the apparatus usually provided at the motor starter.

The fusing factor = Minimum fusing current/ Current rating

There are three standards

Class P Fuses protect against relatively small but sustained overloads with fusing factor of 1.25 (25% overload rating)

Class Q Fuses where protection against relatively small overcurrents is not required, with a fuse factor not exceeding 1.5 for cartridge and 1.8 for semi-enclosed fuses. Motor overload protection to back up motor starter protection Class R Fuses require for protection against relatively large overcurrents (e.g. short circuit protection) 3 x Full load current

PREFERENTIAL TRIPPING

It is essential to prevent interruption of services necessary to maintain propulsion and navigation. These must be safeguarded even if the other services such as domestic supplies are temporarily sacrificed.

There are two ways to safeguard these services. First there must be at least two generators, the rating of which must be such that essential services can be maintained if one set is out of commission. Secondly, a protection must be provided that if sea load is too much for one generator a system of preferential selection will operate.

In some cases the non essential load is relatively too small to warrant additional switchgear. It is generally in larger installations where loads not under direct control of the engineer that they must be fitted. If the heating, lighting and galley were all switched on without prior warning, then the generators could become overloaded.

Without preferential trips this may so overload the generators as to cause a complete shutdown. Therefore non essential services are fed through one or more circuit breakers fitted with shunt retaining coils or shunt tripping coils. Over current relays with time lags are provided for each generator. When overloaded, appropriate relays operate and trip out the non essential services. Some being more important than others, degrees of preference may be given.

SETTING

Usual setting is 150% (50% overload) with a time delay of 15 seconds for generator overload protection and the following times come into operation when the generator reaches 110%.

First tripping circuit 5 seconds
Second tripping circuit 10 seconds
Third tripping circuit 15 seconds

6

Marine Switchboards and Switchgear

MARINE SWITCHBOARD SUPPLIER TO THE DEFENCE AND OIL & GAS INDUSTRY

NCE have designed and built Marine class switchboards and Control Systems for Navy ships, drilling rigs, AHT supply vessels and shore power panels. We provide complete mechanical and electrical engineering design and custom solutions for every application. Every NCE Marine class panel is built to rigorous standards and fully inspected and tested to ensure customer satisfaction. From design to build, NCE is a proven supplier, whether your project is new construction or retrofit, NCE's expert mechanical and electrical design team have years of experience to bring to your application.

SECTORS

- Building services
- Data centres
- Marine / defence
- Education
- Energy
- Health
- Food and beverage
- Leisure & travel
- Industry
- Retrofit

NCE manufactured the 690V switchboards for the MOSPower II AHT supply vessel pictured below. The switchboards contained 6000A Busbars with a fault level up to 120kA. The switchboard also had NCE Switchgear's arc contained Busbar system built in for operator safety. NCE Switchgear supplied the 690V switchboards to MOSPower II - CLICK HERE FOR MOSPOWER II SPECIFICATIONS

NCE offers a broad range of equipment for low-voltage marine applications including:

- Generator Control Panels

- Distribution Switchboards & Switchgear
- Shore Power Panels
- Air Circuit Breaker Replacement
- Main Bus Bar Replacement/ Modifications
- Drillers Control Consoles
- Marine Upgrades

Our major strength is our adaptability to design and configure our manufactured products to produce a complete custom system for your unique project. We further enhance our system packages by customising a variety of components from leading equipment manufacturers including Terasaki, ABB and Schneider.

EASE OF MAINTENANCE

Front mounted components and plug-in circuit breakers assure maintenance accessibility

CUSTOM DESIGN ENGINEERING

We adapt to your application with custom designs and innovative problem solving. NCE designs are thorough and precise using Radan CAD/CAM drafting software.

SAFETY

Operator and equipment safety are prime considerations in NCE design. We adopt Arc Fault Containment in all our Marine Switchgear which further illustrates our goal to supply the best and safest Switchgear to our customers.

MAIN SAFETY DEVICES FOR MAIN SWITCH BOARD ON SHIP?

The Main switch board is an intermediate installation in the ship's power distribution circuit connecting the power generators and power consumers. The power generators on ships are auxiliary engines with alternators and the consumers are different engine room machineries such as motors, blowers etc. It is very important to isolate any type of fault in an electrical system supplied from the main switch board (MSB) or else it will affect all the other system connected to the same. If such isolation is not provided then even a short circuit in a small system can cause blackout of the whole ship.

Therefore, different safety devices are used on board ship and installed on the main switch board (MSB) and electrical distribution panels. This ensures safe and efficient running of machineries and safety of the personal from electric shock even when one system is at fault.

The Important safety devices fitted on main switch board are:

- *Circuit breakers:* A circuit breaker is an auto shut down device which activates during an abnormality in the electrical circuit. Especially during overloading or short circuit, the circuit breaker opens the supplied circuit from MSB and thus protects the same. Different circuit breakers are strategically installed at various locations.

- *Fuses:* Fuses are mainly used for short circuit protection and comes in various ratings. If the current passing through the circuit exceeds the safe value, the fuse material melts and isolates the MSB from the default system. Normally fuses are used with 1.5 times of full load current.
- *Over current relay:* OCR is used mainly on the local panel and MSB for protection from high current. They are installed where a low power signal is a controller. Normally relays are set equivalent to full load current with time delay.
- *Dead front panel:* It is another safety device provided on the Main switch board individual panels wherein you cannot open the panel until the power of that panel is switched off.

Apart from this, maintenance and operational safety plays an important part for the safety of main switch board.

MAIN SWITCH BOARD (MSB) SAFETIES, PROTECTION AND MAINTENANCE

1. Main Switch Board or MSB to be installed on a raised platform.
2. The location where Switch Board or MSB is installed should have good ventilation, illumination and affected by minimum vibration.
3. No water or oil pipes should pass above the Switch Board or MSB.
4. There should be adequate space around allowing comfortable operation ans maintenance.
5. The operating floor space in front and behind the Switch Board or MSB should be covered with anti skid, insulated rubber mats with a minimum thickness of 15 mm.
6. Panel doors should be in locked position.
7. Door opening mechanism should be interlocked with power supply.
8. The enclosure should be drip proof.
9. MSB should be dead front type. (i.e. All the live conductor should pass behind MSB)
10. Every equipment onboard must be properly earthed. Also MSB must be properly earthed.
11. Every incoming and outgoing feed must be protected by a switch fuse or circuit breaker.
12. Insulation resistance should not be less than 1 Mega Ohm in any case. Insulation Resistance (IR) to be checked and records are maintained.
13. All meters must be calibrated periodically and records are maintained.
14. All the meters, switches, fuses, push button, indicating lamps, etc.

should be in working condition and identified with individual name plate or tag.

15. All the interlocks must be in working condition.
16. Operating switch gear, tools such as circuit breaker handle, fuse puller, lamp grip, etc. must be readily available and to be used.
17. Red tag/ name plate indicating high voltage and danger sign should be fitted
18. Appropriate ready to use fire extinguishers should be positioned nearby. ($CO_2$and DCP)
19. Charts indicating drill procedures in case of electric shocks must be displayed near MSB.
20. Wiring diagram laminated or framed indicating supply system must be displayed nearby.
21. All the appropriate spares should be properly preserved and protected, individually packed and identified, and should be readily available for use.
22. Earth fault indication should be fitted and displayed in front panel.
23. 24 volt battery supply emergency light should be provided near MSB.

CONSTRUCTION OF A MAIN SWITCH BOARD (MSB) USED IN SHIPS

A main switch board is an integrated assembly of switching controls, regulator equipment, measuring instruments, indicators and protective devices, and is a necessary apparatus for efficient centralized control of the electrical systems and power operated equipment in ships.

This main switch board distributes electrical power to the electrical loads in various locations in the ship. Should a fault occur at a place, an appropriate protective device in the switchboard will operate to disconnect the faulted circuit. One or more generators are operated and controlled, depending on the size of load, from this main switch board, and supply the power to the feeder circuits through the respective feeder panels.

CONSTRUCTION

A dead front type switch board have no live parts on the front of the panel. Each frame work is rigidly constructed of steel angles for vibration resistance and to which a door type panel or panels of 2.3 mm thick sheet steel are hinged at one side with the other side latched for securing it or them in place. Where inspection or servicing of the internal parts is required, turn each knob latch 90 degrees anticlockwise to unfasten, then the panel is opened and held open by means of a metal bar. Internal devices (instrument transformers, relays, etc.) are so mounted that they are resistant to vibration and impact, yet easily replaceable from the front or back of the switch board.

Insulating material used is mostly laminated-phenolic and special attention is paid to the configuration for temperature and moisture stability,trouble-free operation and safety.

SHIP AUXILIARY SYSTEMS: MAIN SWITCHBOARD

Ships are self contained units floating at sea and hence there needs to be a source of power generation, more than one infact, as well as an emergency backup apart from the main power system. The alternators connected to the diesel generators generate electricity which is then distributed to the different systems of the ship via the main switchboard.

In this article related to ship auxiliary systems we will learn about the general arrangement of the diesel generators and main switchboard on ships. Please note that there are various types of ships, of different sizes, carrying different types of cargoes and so on. Hence there cannot be any hard and fast rule which covers each ship but on reading this article you would be able to grasp the main concept.

MAIN SWITCHBOARD

There are two different types of power requirements on ships namely the 440V and 220V. Infact most of us, even not related to marine engineering understand this pretty well, especially in Asia where normally the 440V is used for industrial connections and the 220V for domestic connections. This is simply because industrial equipments require more power for their operation than general domestic items. Similarly there are heavy loads on ships like say big pumps, winches, cranes etc which require heavy power, while the accommodation and other related uses are similar to the domestic requirements on land.

The main switchboard does the same function like a power substation on land, that is to distribute electricity to different areas as per the requirements and also provide means to monitor and control the power distribution. There are also starter boards which contain the start, stop buttons for most major electrical motors on the ship.

EMERGENCY BACKUP

Apart from the main power source, there is an emergency back up in the form of batteries as well as an emergency diesel generator. The batteries are sufficient to cover up for the time between the main power failure and the injection of power via the emergency system. Of course in smaller ships, there is not emergency generator and the battery backup is sufficient to keep things under control until the main power is restored. The emergency power is supplied mainly to equipments like the steering gear, bilge pump, fire pump, navigational equipment, alarm systems and so forth. In short these are the systems which are vital for the existence in the absence of anything esel

THE LAYOUT

The schematic layout of the main switchboard on a ship is shown in the figure below. As can be seen there is a dual setup for the port and starboard side respectively to keep the functioning as smooth as possible avoiding any drastic situations. The next two pictures show the layout of the emergency diesel generator and the normal generators on a ship at various floors. Both these images have a dual view namely side view and top view so that you can understand more readily as to where is the exact system located.

CIRCUIT PROTECTION AND POWER DISTRIBUTION

TURNING FEAR INTO KNOWLEDGE AND RESPECT

Are you intimidated by your boat's electrical system? If you are, you're certainly far from alone. Electrical shocks, short circuits and fires are potential calamities that ought to inspire serious respect, and frequently lead to fear. Accordingly, those of us who do not understand basic electrical theory often leave onboard electrical work to the experts, generally a wise solution.

This approach can lead you astray if you're anchored in a remote harbor, sailing offshore, or powering on the Intercoastal Waterway and something goes haywire in your onboard electrics. There may not be an expert you can call, and you may need to perform simple repairs on your own. This West Advisor will give you some of the basics about circuit protection and the distribution and monitoring of electrical power on your boat. We can't make you into an instant electrical expert, but perhaps can provide a start. We also offer a selection of books to help you really learn about your boat's electrical system, such as Nigel Calder's excellent Boatowner's Mechanical and Electrical Manual, among others.

CIRCUIT PROTECTION DEVICES

Wires used in electrical circuits need some method of protection to prevent fire or damage if there is a short circuit or an over-loaded circuit. A short circuit occurs when electricity travels from the positive to negative conductor without passing through an electrical load (a device that consumes the energy, creating electrical impedance), allowing more electricity to flow than the wire can handle. Heat builds up and eventually melts the insulation, which can cause a fire or damage to the electrical system.

Two types of products protect wires from short circuits: fuses and circuit breakers. Fuses are devices that melt or destruct at a certain current flow, like five amps. Once they melt they must be replaced. Circuit breakers are mechanical devices that trip and open the circuit without damage to themselves.

Since circuit breakers can also be used as switches, they are frequently stacked in a row in a distribution panel. They are also measured by the current (amperage) it takes to cause them to trip.

Fuses are also used to protect an electrical device from its own destruction. For example, fuses will commonly blow if you hook up a VHF radio with reverse polarity (positive connected to negative, and vice-versa). The fuse blows to protect the inside of the radio from a meltdown.

STYLES AND SIZES OF FUSES

Fuses are available in a variety of configurations. Slow-blow fuses hesitate for a few seconds before blowing to allow products with a short term, high amperage draw to start up. AGC fuses are small glass cylinders, which allow you to see whether the fuse element is intact or not. They are used in inexpensive fuse panels. ATC fuses are automotive style, and plug into a two-pronged socket. They are rapidly replacing AGC fuses in a number of applications. Fuse holders provide a way to install a fuse in the positive conductor of a circuit. They are available to fit AGC and ATC fuses, and are available in waterproof and non-waterproof versions. Fuses should be installed as close to the beginning of the positive conductor as possible, to protect the greatest amount of wire.

Here is an over-simplified guide to size selection:

- The amperage of the load largely determines everything else.
- This, along with the distance to the load, determines the size of wire based on the wire selection tables.
- The current carrying ability of the wire determines how big a fuse to use.

CIRCUIT BREAKER SIZE SELECTION

Selecting the correct breaker size is a real problem for most custom and standard panels on the market. The issue is the impossibility of anticipating what breakers will be needed on a distribution panel, since all boats are different. Blue Sea Systems solves this problem simply. They reason that the majority of circuits onboard will have loads of less than 15A (certainly that is the case with most electronics, lights, pumps, small refrigerators, etc.). They also reason that most wire used for circuits will be at least 14-gauge, although it may be as small as 16-gauge. Therefore, they supply their panels with 15A circuit breakers throughout the panel, for the following reasons:

- If the circuit uses 16-gauge (small) wire, which has an ampacity of 25A, the breaker will protect it.
- 15A is larger than most loads on a boat.
- A small fuse can be used at the load, or a smaller breaker substituted in the panel if smaller current protection is necessary.

DISTRIBUTION PANELS

Distribution panels let you selectively turn on and off the circuits that provide electricity to DC loads on board. They may be combined with one or more monitoring functions like amperage, voltage, or amp-hours. Most boats over 20' or so will have one or more distribution panels that allow circuits to be switched off and on, and provide some form of over-current protection to protect wires and loads from damage.

Over-current protection means that if something happens to an electrical circuit that makes it consume more electricity than normal, the circuit will automatically turn off. This is done with either circuit breakers or fuses, just like in home wiring systems. Most panels have between 4 and 12 circuits. This means they can control and protect circuits used for a variety of applications like radios, lights, bilge pumps, refrigerators or electronics.

Distribution panels from many vendors come in industry-standard sizes so they can be stacked vertically or horizontally as needs dictate to create larger capacity. Panels are most commonly 5 1/4" wide, with lengths of 3 3/4", 7 1/2", 10 1/2", or 11 1/4".

AC panels (alternating current, or shore power) are almost identical to DC panels (direct current, or battery power), but they include a main breaker to disconnect all the AC circuits at once. This is like the main circuit breaker on the outside of a house. The individual circuit breakers in an AC panel are the same as those used in a DC panel. They also have a reverse polarity light that glows when the dock or inverter is wired incorrectly. Many marinas are wired incorrectly, so that the white conductor is actually hot and the black is neutral. Don't panic about the terms; what it means is you can either electrocute yourself or have severe electrolysis on your boat if the problem is not rectified. If this reverse polarity light goes on, you should disconnect AC power from your boat and find out what the problem is.

SELECTING THE CORRECT PANEL

Choose a panel based on the number of circuits on the boat, for both AC and DC. We also offer panels that have special features like waterproofness, lighting and built-in meters.

DC panels vary based on a number of design details. Here are some things to look for:

- Panels should have backlit labels so you can read the function at night. This is much better than having a small light shining down on the labels from above.
- Circuits should have indicator lamps or LEDs, which show if the circuit is on. This makes it easy to see at a glance what is on and off.
- Since a pair of wires comes from each circuit, a positive and a negative bus should be provided.

- Almost all panels use Carlingswitch breakers, which attach with two screws. The nice way to install them is with countersunk flat head fasteners that are painted black to match the panel.
- The panel size and color should match the panels that came on the your boat, or the panel purchased from us last season.

Not surprisingly, the Blue Sea Systems panels include all the above features. They are absolutely the nicest, most well thought out panels on the market. AC panels are slightly different than DC versions because you need to disconnect both the hot and neutral conductors at one point in the system (the Master AC Breaker). You may also need to select between two or more sources of AC power on a boat: inverter, generator, or shore power. The challenge is that while it is generally OK to have two DC sources on-line at the same time (battery charger, battery and solar panel), it is not OK to have two AC sources on simultaneously. AC sources have to be perfectly synchronous (be positive and negative at the exact same instant) and it becomes difficult to do this with multiple sources. This is why AC Source Selector panels generally have a slider or lock which prevents more than one of the circuit breakers from being turned on at the same time. On the Blue Sea Systems panel, there is a plastic slider that prevents both breakers from being activated at once, which might put the shore power and inverter on simultaneously.

ABYC STANDARDS

The ABYC (American Boat and Yacht Council) has extensive information on how to wire distribution panels, since doing it the wrong way can lead to boat fires and electrocutions.

MONITORS AND CONTROL PANELS

Electrical system monitors provide a way to see and understand what is normally invisible. Since most of us have trouble counting electrons or understanding voltage, monitors are used to show us what is happening in our electrical system. For decades, boaters have gone to sea with only an approximate idea of how much energy their batteries contained, and when it would be appropriate to charge their batteries. It has also been difficult to tell when a boat's batteries needed replacement, and batteries are often replaced prior to the end of their useful life. Therefore, battery monitors or electrical system monitors provide a great benefit.

Electrical system monitors have three functions:

- Voltmeters are used to show the approximate state of charge of batteries, and to show when charging sources are operating. They also allow boaters to adjust regulators, like those connected to a solar panel, so they charge batteries at the correct voltage. To be useful, voltmeters should be able to measure tenths of a volt.

- Ammeters are used to show how much current is flowing, measured in amps. A windlass might draw 150A, while a GPS might draw 0.05A, so ammeters have to operate over a wide range of values. They can also be used to show how rapidly your batteries are being discharged, and how much current your alternator is generating.
- Amp-hour meters display how much energy has been removed from your boat's batteries. They start at 0 Ah, count "down" as energy is removed (-45Ah) and count "up" as batteries are charged.

When we refer to an electrical monitor, we imply that it contains all three basic functions: volts, amps, and amp-hours. Since amp-hours are the best way to know your battery's state of charge, and since voltage and amperage are also helpful, a combination meter which measures all three is the best solution for boats over 30'.

We've seen real advancements in electrical system monitoring in the last decade, and now we have really great, cost-effective combination meters that tell you exactly what you need to know about your electrical system. Besides the three main functions, meters like the LinkPRO tell you things like operating time remaining until you need to recharge, deepest battery discharge, average level of discharge, and charging efficiency. There isn't much to the sizing of a monitor, since almost all of them will read to 500A, enough for all but very large boats.

Separate meters can be added to existing electrical systems. Voltmeters, ammeters, and percent meters by themselves are not as comprehensive, but are still helpful to see the status of your system. Most large inverters can be hooked up to remote panels, which vary from remote ON-OFF switches to complete monitoring systems. Besides turning the inverter or battery charger on and off, many panels monitor shore power breaker settings, battery state-of-charge and equalization, and some automatically start your generator when the battery voltage drops or cabin temperature rises, so your air conditioner can run.

HIGH VOLTAGE SWITCHGEAR

Click to show/hide page index

The power system deals with voltage above 36KV, is referred as high voltage. As the voltage level is high the arcing produced during switching operation is also very high. So, special care to be taken during designing of high voltage switchgear. High voltage circuit breaker, is the main component of HV switchgear, hence high voltage circuit breaker should have special features for safe and reliable operation. Faulty tripping and switching operation of high voltage circuit are very rear. Most of the time these circuit breakers remain, at ON condition, and may be operated after a long period of time. So CBs must be reliable enough to ensure safe operation, as when required. High voltage circuit breaker technology has changed radically in the last 15 years. Minimum oil circuit breaker, air blast circuit breaker and SF_6 circuit breaker are mostly used for high voltage switchgear. Vacuum circuit breaker is rarely used for this purpose as till date vacuum technology is not adequate for interrupting very high voltage short circuit current. There are two types of SF_6 circuit breaker, single pressure SF_6 breaker and two pressures circuit breaker.

The single pressure system is the state of art for high voltage switchgear system, in present time.

Now days SF_6 gas as arc quenching medium, has become most popular for high and extra high voltage electrical power system. Although, SF_6 gas has strong impact on the greenhouse effect. It has 23 times stronger impact on the greenhouse effect, than that of CO_2. Hence, leakage of SF_6 gas during the service life of circuit breaker must be prevented. In order to minimize the emission of SF_6 gas, the N_2 - SF_6 and CF_4 - SF_6 gas mixture, may be used in circuit breaker in future, as substitute of pure SF_6. It must always be taken care of that, no SF_6 gas comes out in atmosphere during maintenance of the CB. On the other hand, SF_6 circuit breaker has the major advantage of low maintenance.

High voltage switchgears are categorized as,

1) Gas insulated indoor type, 2) Air insulated outdoor type.

Again, outdoor type air insulated circuit breakers are classified as,

1) Dead tank type circuit breaker 2) Live tank type circuit breaker

In dead tank type CB, the switching device is located, with suitable insulator supports inside a metallic vessel(s) at ground potential filled with insulating medium.

In live tank circuit breaker, the interrupts are located in an insulated busing, at the system potential. The live tank circuit breakers are cheaper and required less mounting space. There are mainly three types of circuit breaker, as we said earlier, used in high voltage switchgear system i.e. air blast circuit breaker, SF_6 circuit breaker, oil circuit breaker. And vacuum circuit breaker is rarely used.

AIR BLAST CIRCUIT BREAKER

In this design, a blast of high pressure compressed air is used to quench arc between two detaching contacts, when the arc column ionization is least at currents zero.

OIL CIRCUIT BREAKER

This is further classified as bulk oil circuit breaker (BOCB) and minimum oil circuit breaker (MOCB). In BOCB, the interrupting unit is placed inside an oil tank of earth potential. Here oil is used as both insulating and interrupting medium. In MOCB on the other hand, the oil requirement can be minimized by placing the interrupting units in an insulating chamber at live potential on an insulator column.

SF_6 CIRCUIT BREAKER

SF_6 gas is widely used as arc quenching medium in HV applications today. Sulfur hexafluoride gas is a high electronegative gas having excellent dielectric and arc quenching properties. High dielectric and insulating properties of SF_6, make it possible to design high voltage circuit breaker with smaller overall dimension, shorter contact gap. Excellent insulating property helps to design and construct indoor type high voltage switchgear.

VACUUM CIRCUIT BREAKER

In vacuum, there is no further ionization between two separated current carrying contacts, after current zero. The initial arc is caused by it will die as soon as next zero crossing but as there is no provision of further ionization once the current is crossed its first zero, the arc quenching is completed. Although the arc quenching method is very fast in VCB, but till it is not a suitable solution for high voltage switchgear, as VCB made for very high voltage level is not economical at all.

The essential features to be provide in high voltage circuit breaker, to ensure safe and reliable operation the breakers used in high voltage switchgear, must be capable of being operated safely for,

1) Terminal faults. 2) Short line faults. 3) Transformer or reactors magnetizing current. 4) Energizing long transmission line. 5) Charging capacitor bank. 6) Switching of out of phase sequence.

TERMINAL FAULT

Generally the load connected to the power system is inductive in nature. Due to this inductance, when short circuit current is just interrupted by a circuit breaker, there is a chance of high restriking voltage of high frequency oscillation in order of few hundred Hz. This voltage has two parts 1) Transient recovery voltage with high frequency oscillation immediately after the arc extinction. 2) After die down this high frequency oscillation, power frequency recover voltage appears across the CB contacts.

TRANSIENT RECOVERY VOLTAGE

Just after extinction of arc transient recovery voltage appears across the CB contacts, with high frequency. This transient recovery voltage ultimately approaches to open circuit voltage. This recovery voltage can be represented as

The frequency of oscillation is governed by the circuit parameter L and C.

The resistance present in the power circuit damps out this transient voltage. The transient recovery voltage has not a single frequency, it is combination of may different frequencies due to complexity of the power network.

POWER FREQUENCY RECOVERY VOLTAGE

This is nothing but open circuit voltage appears across the CB contacts, just after the transient recovery voltage damped out. In three phase system the power frequency recovery voltage differs in different phase. It is highest in first phase.

If the network neutral is not earthed, the voltage across the first pole to be cleared is 1.5U where U is the phase voltage. In an earthed neutral system it will be 1.3U. By using damping resistor, the magnitude and rate of rise of transient recovery voltage can be limited. The dielectric recovery of the arc quenching medium and rate of rise of transient recovery voltage has great influence on the performance of the circuit breaker used in high voltage switchgear system.

In air blast circuit breaker, once ionized air is deionized very slowly, thus air takes long time to recover dielectric strength. That is why it is preferable to used low value breaker resistor to slow down rate of rise of recovery voltage. On the other hand ABCB is less sensitive to the initial recovery voltage because of high arc voltage in SF_6 circuit breaker, the interrupting medium (SF_6) has faster rate of recovery of dielectric strength, than air. Lower arc voltage, makes SF_6 CB more sensitive to the initial recovery voltage.

In oil circuit breaker, during arc having pressurized hydrogen gas (produced during recombination of oil due to arc temperature) provides quick recovery of dielectric strength immediate after current zero. Hence OCB is more sensitive to rate of rise of recovery voltage. It is also more sensitive to initial transient recovery voltage.

SHORT LINE FAULT

Short line fault in transmission network is defined as the short circuit faults occurred, within 5km of the line length. Double frequency being impressed on the circuit breaker and the difference of source and line side transient recovery voltage. Both voltages start from instantaneous values at the opposition of the circuit breakers prior to the interruption. On the supply side the voltage will oscillate at supply frequency and ultimately approaches to open circuit voltage. On the line side, after interruption, trapped charges initial traveling waves through the transmission line, since there is no driving voltage on the driving side, the voltage ultimately becomes zero because of the line losses.

HIGH VOLTAGE (AC) SYSTEMS

No one who is not trained in marine high voltage systems should ever perform wiring on a boat. The potential for creating deadly system faults that are injurious to both people and the vessel itself is too risky. That includes something as apparently simple as wiring a new service outlet, for if you don't understand the principles of the system, you are likely to unknowingly create a problem. Just because you know how to do wiring around the house, doesn't mean you know what you're doing on a boat.

GROUNDS AND GROUNDING

One of the least understood aspects of a boats electrical system, and the most troublesome, is the proper method of grounding. That we often get questions of whether AC or DC electrical equipment should be grounded to the boat's bonding system is illustrative of this point. AC and DC grounding systems are two separate systems, for distinctly different reasons. If you don't understand these systems, you run the distinct risk of creating a disaster. Actually, there are four separate ground systems: DC ground, AC ground, AC grounding (or bond), and the vessel's bonding system. You can add to this lightning and HF radio grounds as well. Do you know the principles of each? Are you sufficiently confused to discourage you from doing your own wiring? I hope so. For unless you understand each thoroughly, you're headed for trouble.

The AC ground and grounding systems are "free floating," meaning that they do not ground on the vessel, but only to shore. The ground, or neutral, is a current carrying conductor, and is the source of many troubles because people do not regard it as such. The grounding, bond or green wire is the "safety" intended to channel current safely to ground in the event of a short circuit. Both of these circuits are capable of conducting current and can be the source of electrolysis when there are system faults with the dock or marina wiring. This is very easy to test for.

There is only one point where the DC side is grounded, and that is at the battery. It, too, is a "free floating" system in which nothing is ever grounded to any metallic part of the vessel, most especially not the bonding system. Just like a car sitting on rubber tires, completely insulated from earth potential, the battery itself provides the negative potential. The bonding system, also green wire, has nothing to do with electrical systems. Underwater metals are simply wired together to equalize differences in potential of different kinds of metal. Nothing should ever be grounded to the bonding system. Unfortunately, some people don't understand this and use it to ground electrical equipment, occasionally with disastrous results.

BONDING SYSTEMS

Bonding simply means wiring all the boats underwater metals together. This is done because of the galvanism caused by the different metals. By wiring them together, the differing potentials are equalized. Bonding does not solve problems of galvanism or electrolysis, but it does spread the the flow of current around over more metal, so that 1/4 volt or so won't cause any damage. Whereas if that 1/4 volt were going to one small seacock, it would probably eat it up in a hurry. In other words, bonding lessens the effect of small amounts of current. On the other hand, it also spreads it around to all underwater metals so that higher currents end up damaging everything.

Bonding systems use wire and ordinary crimped ring terminals. After a while these get wet and corroded. Electricity doesn't flow very well through corroded metal, so your bonding system after a while stops working. To maintain it, simply cut off the old terminals and install new ones. Do you have wires attached to sea cocks with hose clamps? Forget it. This is putting stainless and copper together, which are galvanically incompatible and it won't work.

YOUR BOTTOM PAINT

What does bottom paint have to do with electrical systems? Nowadays, with copper based paints, a lot. If, the next time your boat is hauled and you see large ugly burn patterns around all your underwater metals, you got a stray current problem. Copper-based bottom paints react severely to stray current, and serves as a great indicator. Sort of litmus paper for electrical problems.

Of course, the common wisdom is that the stray current "is from the marina." Or it's always the other guy's boat that is causing your problem. Don't bet on it. Most stray current problems are sourced on the boat in which they appear. Otherwise, everybody in the marina would have the same problem.

ELECTROLYSIS AND GALVANISM

Electrolysis is a word that is badly abused by boaters who don't really know what it means, so let me correct this right now. First, understand that all boats have an electrical potential. That's because of all the different metals on the boat which, themselves have differing electrical potentials. This is exactly the same principle that makes a dry cell battery generate electricity. This electrical potential is called galvanism and is the reason why we put zincs on boats. Electrolysis is stray current escaping from the system and is most damaging. It is an abnormal condition. When this happens, it will eat up the zincs in no time, usually leaving that metal looking bright and shiny. Therefore: Shiny zincs = electrolysis. Dull eroded zincs = galvanism.

I spent two years putting a meter on every boat that was hauled for survey. The average boat generates about 1/4 volt DC current and going as high as 1/3 volt without causing damage. But when it gets up to 1/2 volt, you got a problem. Zincs will erode rapidly and underwater metals begin to be affected.

SHORE POWER CORDS

The single largest cause of problems with shore power systems results from failure to maintain the connectors on both the cord and the boat connectors.

These devices are exposed to water and over time suffer from corrosion and general wear. High resistance caused by corroded, bent or worn connectors results in high resistance which causes overheating, which further amplifies the power drop.

This not only creates conditions for a potential fire, but causes electrical equipment to work harder, resulting in reduced life span of equipment. IT PAYS TO MAINTAIN SHORE POWER CONNECTIONS.

You can perform a very simple check just by placing your hand on the shore cord near the connection to determine if it is heating up. Obviously, this should be done while you have a lot of equipment turned on. If it's anything but slightly warm, not more than 110 degrees, suspect a problem. Shore power connectors should be dismantled at least once per year, cleaned and repaired as necessary. Most of these connectors have replaceable parts. If you drop your shorepower connector in the water, you must take it apart, clean and dry it. Otherwise, expect it to burn up.

We recommend that you buy only the highest quality power cords, as these will last longer and have the advantage of replaceable connector parts. Cheap connectors usually can't be taken apart. We also advise against ever using the three-pronger household type adapters as this type of connector is highly unreliable and prone to causing system faults and fires. Only the twist-lock type connector is suitable. One more thing: If you are not turning off the dock breaker before disconnecting the power cord, start doing it now. Not only do you risk getting electrocuted, but disconnecting an energized connector damages the contacts. Also consider what happens if you drop the energized cord in the drink!

POLARITY

This is an issue only with 125 VAC systems since 250 VAC systems will not function with wrong polarity. Since you have three terminals on a shore connection, wrong polarity can mean that any of these wires are in the wrong position. Not only should you pay attention to the polarity indicator on your boat, we recommend that you keep a plug in polarity indicator aboard and use it every time you hook up to shore power at a different location.

Reverse polarity is not only an electrocution hazard, but can also damage electrical equipment. It is most often found with the three prong spade connectors (household type), but occasionally twist lock connectors as well, particularly in marinas with dilapidated equipment.

Never trust the power supply at strange docks, but always check the polarity. When hooking up to strange docks, always check your volt meters to make sure you have adequate voltage. Low voltage is very damaging to electrical equipment. Turn on the stove or water heater and watch what happens to the meter.

MAIN CIRCUIT PROTECTION

Many people think that the circuit breakers on the dock protect their boat. They do not; they only protect the dock wiring. Your main circuit breaker protects your boat's systems. But what about that section of wiring and connectors between your main panel and the dock breaker? Well, the fact is that it is unprotected. which is why so many fires occur. Check out all the top end boats and you will find that they have circuit protection located directly at the shore connectors. Which is why we recommend that you should too. Having slow blow cartridge fuses installed directly at the connectors can go a long way toward preventing fires and burned up shore cords, particularly if you are a traveler and frequently rely on uncertain power supplies. Circuit breakers should NEVER be installed on the exterior of the boat. Only gasketed, water proof cartridge holders should be used.

CIRCUIT BREAKERS

Circuit breakers wear out, and when they do they work less well, or not at all. If you are using circuit breakers as ON/OFF switches, you are helping them wear out that much faster. It also damages breakers when you shut off equipment via the breaker. This causes arcing at the contact points which damages the points. When connecting and disconnecting shore power, you should always turn OFF equipment at the appropriate switch on the equipment. Then shut the main breaker off. Do not ever simply throw the main breaker off to shut down equipment that is operating. The circuit breaker arcs and damages it.

Also be aware that any equipment run by a motor, such as air conditioning and refrigeration equipment, start up with an initially much higher amperage than the normal running amperage. An air conditioner that runs at 14 amps may have a start up amperage of 20 amps, so that if you just go and turn all the equipment on at once, it overloads the system. Then the circuit breaker gets hot and won't stay engaged until it cools down. Ergo, start up heavy equipment one item at a time, allowing it time to cycle into its normal operating voltage before turning something else on. For example, don't turn the AC, refrigerator and icemaker all on at once and not expect the breaker to pop.

CHRONIC BREAKER POPPING

It's human nature that when a problem appears, we wish it to go away. Breakers that pop frequently are signaling that there is a problem, which could either be the breaker, or something in the circuit. Yet most people will keep on attempting to make the breaker engage. This can be dangerous because you may cause the contact points of the breaker to fuse together from arcing, in which case it will never trip again. The above photo shows what can happen when you do this. DO NOT ATTEMPT TO ENGAGE A BREAKER THAT IS OVERHEATED BY FORCING IT. You must allow it to cool down.

If you are experiencing chronic problems with circuit breakers popping, first check how much current draw is involved. A single 30 amp circuit is not much when you're running things like air conditioners, water heaters and battery chargers. One very simple way to check whether you're dealing with an overload problem is to add up the amperage draw of each piece of equipment. List both the start up and run amperages. You will usually find the amperage given right on the equipment label. By making a list of the total power demand, you'll get a good idea of what you can and cannot operate simultaneously, particularly when starting the equipment. If you have an ammeter on your panel, check it against the amperage tally you made. Ideally, you should try to hold power consumption at 80% or less than the line rating.

Check the breaker by allowing it one hour (or whatever it takes) to cool down. Turn the equipment off and, after it is cool, reengage the breaker. Now turn the equipment back on. Place your finger on the front of the breaker and note its temperature. If it does not heat back up again, then the problem was probably a start-up overload. If the temperature rises again, there is a fault in the circuit or the breaker.

7

High Voltage in Ships

We all know about the voltages used on board a ship. It is usually a 3phase, 60Hz, 440 Volts supply being generated and distributed on board. Every day the owners and designers aim for bigger ships for more profitability. As the ship size increases, there is a need to install more powerful engines and other machineries. This increase in size of machineries and other equipment demands more electrical power and thus it is required to use higher voltages on board a ship.

Any Voltage used on board a ship if less than 1kV

(1000 V) then it is called as LV (Low Voltage) system and any voltage above 1kV is termed as High Voltage. Typical Marine HV systems operate usually at 3.3kV or 6.6kV. Passenger Liners like QE2 operate at 10kV.

DEFINATION OF HV:

The numerical definition of high voltage depends on context. Two factors considered in classifying a voltage as "high voltage" are the possibility of causing a spark in air, and the danger of electric shock by contact or proximity. The definitions may refer to the voltage between two conductors of a system, or between any conductor and ground.

In electric power transmission engineering, HIGH VOLTAGE is usually considered any voltage over approximately 33,000 volts. This classification is based on the design of apparatus and insulation. The International Electro technical Commission and its national counterparts (IET, IEEE, VDE, etc.) define high voltage as above 1000 V for alternating current, and at least 1500 V for direct current—and distinguish it from low voltage (50–1000 V AC or 120–1500 V DC) and extra-low voltage (<50 V AC or <120 V DC) circuits. This is in the context of building wiring and the safety of electrical apparatus.

- In the United States 2005 National Electrical Code (NEC), high voltage is any voltage over 600 V (article 490.2).
- British Standard BS 7671:2008 defines high voltage as any voltage difference between conductors that is higher than 1000 V AC or 1500 V ripple-free DC, or any voltage difference between a conductor and Earth that is higher than 600 V AC or 900 V ripple-free DC.

WHAT IS CLASSED AS HIGH VOLTAGE?

In marine practice,

- Voltages below 1,000Vac (1kV) are considered low voltage, and
- High voltage is any voltage above 1kV. Typical marine high voltage system voltages are 3.3kV, 6.6kV and 11kV.

THE MAJOR DIFFERENCES BETWEEN HIGH VOLTAGE SUPPLY AND LOW VOLTAGE SUPPLY ON BOARD SHIPS ARE:

1. High voltage systems are more extensive with complex networks and connections,
2. Isolated equipment MUST BE earthed down
3. Access to high voltage areas should be strictly limited and controlled
4. Isolation procedures are more involved
5. Switching strategies should be formulated and recorded
6. Specific high voltage test probes and instruments must be used
7. Diagnostic insulation resistance testing is necessary
8. High voltage systems are usually earthed neutral and use current limiting resistors
9. Special high voltage circuit breakers have to be installed

WHY HIGH VOLTAGE IN SHIPS?

- Higher power requirements on board vessels is the foremost reason for the evolution of HV in ships.
- Higher power requirements have been necessitated by development of larger vessels required for container transport particularly reefer containers.
- Gas carriers needing extensive cargo cooling Electrical Propulsion.
- For ships with a large electrical power demand it is necessary to utilise the benefits of a high voltage (HV) installation.
- The design benefits relate to the simple ohms law relationship that current (for a given power) is reduced as the voltage is increased. Working at high voltage significantly reduces the relative overall size and weight of electrical power equipment.

AS PER OHMS LAW
POWER = VOLTAGE x CURRENT
For a given Power,
Higher the Voltage, Lesser is the Current
440 KW = 440,000 Watts
= 440 Volts × 1000 Amps
=1100 Volts × 400 Amps
=11000 Volts × 40 Amps

- When large loads are connected to the LV system the magnitude of current flow becomes too large resulting in overheating due to high iron and copper losses.

$P = VI \cos\Phi$

Copper loss $= I^2 R$ [kW]

HV levels of 3.3 kV, 6.6 kV and 11 kV are regularly employed ashore for regional power distribution and industrial motor drives. For example, a motor (let us assume a bow thruster), may be of a smaller size if it designed to operate on 6600 Volts.

For the same power, the motor would be of a smaller size if it is designed for 6600Volts when compared to 440Volts. Thus these are the major reasons why recent ships have shifted towards high voltage systems. The main disadvantage perceived by the user/maintainer, when working in an HV installation, is the very necessary adherence to stringent safety procedures.

ADVANTAGES/DISADVANTAGES OF USING HV ADVANTAGES:

Advantages:

For a given power, Higher voltage means Lower current, resulting in:

- Reduction in size of generators, motors, cables etc.
- Saving of Space and weight
- Ease of Installation
- Reduction in cost of Installation
- Lower losses – more efficient utilization of generated power
- Reduction in short circuit levels in the system which decides the design and application of the electrical equipment used in the power system.

Disadvantages:

1. Higher Insulation Requirements for cables and equipment used in the system.
2. Higher risk factor and the necessity for strict adherence to stringent safety procedures.

Marine Electrical System

- Some installations may feed the ships sub stations directly with HV and step- down to 440 V locally.
- The PEM drives in this example are synchronous motors which require a controlled low voltage excitation supply current to magnetise the rotor poles.
- This supply is obtained from the HV switchboard via a step-down transformer but an alternative arrangement would be to obtain the excitation supply from the 440 V ER sub switchboard.

- Hazardous Electrical Voltage Training ChecklistThe training requirements below apply to all employees who face a risk of electrical shock that is not reduced to a safe level by electrical installation requirements and
- who must work on or near energized components.

All Qualified High Voltage Electrical Workers who work on high voltage equipment (> 600 volts) are required to be trained on safety-related work practices that pertain to their jobs and in the following topics below: The skills and techniques necessary to distinguish exposed live parts from other parts of electrical equipment.

- The skills and techniques necessary to determine the nominal voltage of exposed live parts.
- The clearance distances and the corresponding voltage to which the Qualified Person will be exposed.
- Safely de-energizing of parts and subsequent electrical lockout and tagging procedures as required by the electrical standard.
- Proper precautionary work techniques.
- Proper use of PPE to include non-conductive gloves, aprons, head protection, safety glasses, and face shields.
- Proper selection and use of rated test instruments and equipment, including the capability to visually inspect all parts of the test equipment for defects.
- Use of insulating and shielding materials for employee protection to include auxiliary shields, guards, mats, or other specific equipment.
- Proper use of insulated tools or other non-conductive devices such as fuse pullers, fish tapes, hot sticks, ropes, or handlines.
- The importance of illumination and to work only in properly illuminated areas.
- Proper work techniques for work in enclosed or confined work spaces.
- Removal or special handing of any conductive materials and equipment.
- Proper and safe use of portable ladders around electrical equipment.
- Removal of any conductive jewelry or apparel.
- Proper alerting techniques such as using safety signs and tags, barricades,attendants, and work practices.
- Any other safety related work practice not listed above but necessary for them to safely do their job

ELECTRIC SHOCK:

Voltages greater than 50 v applied across dry unbroken human skin can cause heart fibrillation if they produce electric currents in body tissues that happen to pass through the chest area. Accidental contact with high voltage supplying sufficient energy may result in severe injury or death. This can occur as a person's body provides a path for current flow, causing tissue damage and heart failure. Other injuries can include burns from the arc generated by the accidental contact. These burns can be especially dangerous if the victim's airways are affected.

HAZARDS OF HIGH VOLTAGE

Arcing:

An unintentional electric arc occurs during opening of a breaker, contactor or switch, when the circuit tries to maintain itself in the form of an arc. During an insulation failure, when current flows to ground or any other short circuit path in the form of accidental tool slipping between conducting surfaces, causing a short circuit.

Results of an Electric Arc:

Temperatures at the arc terminals can reach or exceed 35,000° f or 20,000°c or four times the temperature of sun's surface. The heat and intense light at the point of arc is called the arc flash. Air surrounding the arc is instantly heated and the conductors are vaporised causing a pressure wave termed as ARC BLAST.

Hazards of an Arc Flash:

- During an arc flash, sudden release of large amounts of heat and light energy takes place at the point of arc.
- Exposure frequently results in a variety of serious injuries and may even be fatal, even when the worker is ten feet or more from the arc center.
- Equipments can suffer permanent damage.
- Nearby inflammable materials may be ignited resulting in secondary fires.

Hazards of Arc Blasts & ejected materials:

- An arc flash may be accompanied by an arc blast
- The arc blast causes equipment to literally explode ejecting parts with life threatening force. – Heated and vaporised conducting materials surrounding the arc expand rapidly causing effects comparable to an explosive charge.
- They may project molten particles causing eye injuries. The sound that ensues can harm the hearing.

- Potential injuries:
- At some distance from the arc, temperatures are often high enough to instantly destroy skin and tissue. Skin temperatures above 100°C

(about 210°F) for 0.1sec result in irreversible tissue damage, defined as an incurable burn.

- Heated air and molten materials from arc faults cause ordinary clothing to burst into flames even if not directly in contact with the arc. Synthetic fibers may melt and adhere to the skin resulting in secondary burns.
- Even when safety goggles are worn, arc flash may cause severe damage to vision and or blindness. Intense UV light created by arc flash can damage the retina. Pressure created from arc blasts can also compress the eye, severely damaging vision.
- Hearing can also be affected by the loud noise and extreme pressure changes created by arc blasts. Sound blasts with arc blasts exceed 140dB which is equal to an airplane taking off. Sudden pressure changes exceeding 720lbs/sq.ft for 400ms can also rupture eardrums. Even at lesser pressure, serious or permanent damage to hearing may occur.

SHORT CIRCUIT

A short circuit (or a fault) is said to have taken place when the current is not confined to its normal path of flow but diverted through alternate path(s).

- During short circuit, the current rises much above the normal value.
- Short circuit level is the maximum possible current that flows at the point of fault during a short circuit.

Effects of Short Circuit:

High currents during Short circuits can cause damage to electrical installation by giving rise to excessive Thermal Stresses, Mechanical Stresses, Arcing.

Methods adopted to prevent effects of short circuit in a system:

- A well-designed Protective Relay system trips out a breaker(s) and isolates the faulty circuit from the power source within a short time to prevent/minimise effects of high short circuit current, as and when it occurs.
- The equipment in the system, the cables, the switchgear, the busbar, the generators are designed to withstand the effects of short circuit during that short period.

Calculation of the short circuit levels in the system is therefore required to help in:

a. Designing an appropriate Protective Relay System
b. Choosing the right switchgear with suitable short circuit withstand capacity to be used in the system.

HIGH VOLTAGE SAFETY AND PRECAUTIONS

Making personal contact with any electric voltage is potentially dangerous. At high voltage (>1000 V) levels the electric shock potential is lethal. Body resistance decreases with increased voltage level which enhances the current flow. Remember that an electric shock current as low as 15 mA can be fatal. So,the risk to people working in HV areas is greatly minimised by the diligent application of sensible general and company safety regulations and procedures.

Personnel who are required to routinely test and maintain HV equipment should betrained in the necessary practical safety procedures and certified as qualified for this duty. Approved safety clothing, footwear, eye protection and hard hat should be used where danger may arise from arcs, hot surfaces and high voltage etc. Safety equipment should be used by electrical workers includes insulated rubber gloves and mats. These protect the user from electric shock.

Safety equipment is tested regularly to ensure it is still protecting the user. Testing companies can test at up 300,000 volts and offer services from glove testing to Elevated Working Platform or EWP Truck testing. A insulated material or rubber mat can be used as a dead front of all electrical installations and equipments. The access to HV switchboards and equipment must be strictly controlled by using a permit-to-work scheme and isolation procedures together with live-line tests and earthing-down before any work is started. The electrical permit requirements and procedures are similar to permits used to control access in any hot-work situation, e.g. welding, cutting, burning etc. in a potentially hazardous area.

HIGH VOLTAGE SAFETY RULES AND PROCEDURES

All safety rules presented in this document are intended to ensure safe working conditions while working with potentially dangerous voltages. It is assumed that all personnel working with potentially dangerous voltages have been trained in basic electrical safety procedures.

1. This guidance does not apply where equipment has been isolated, discharged, disconnected and removed from the system or installation.
2. Equipment that is considered by an Authorised Person (HV) to be in a dangerous condition should be isolated elsewhere and action taken to prevent it from being reconnected to the electricity supply.
3. All working on, or testing of, high voltage equipment connected to a system should be authorised by a permit-to-work or a sanction-for-test following the procedures as described in Practical Exercises no. 4
4. No hand or tool (unless the tool has been designed for the purpose) must make contact with any high voltage conductor unless that

conductor has been confirmed dead by an Authorised Person (HV) in the presence of the Competent Person (HV).

5. Where any work or test requires an Accompanying Safety Person (HV) to be present, he/she should be appointed before that work or testing can begin.
6. Voltage test indicators should be tested immediately before and after use against a test supply designed for the purpose.
7. Where the procedures involve the application of circuit main earths, the unauthorised removal of such earths should be prevented, wherever practicable, by the application of safety locks.
8. Where the procedures involve the removal of circuit main earths, that is, testing under a sanction-for-test, the earths will be secured with working locks. The keys to these locks will be retained by the Duty Authorised Person (HV), who will remove and replace the earths as requested.

Precaution prior to live voltage and phasing checks:

1. Where live phasing is to be undertaken, the area containing exposed live conductors should be regarded as a high voltage test enclosure.
2. Approved equipment used for live voltage and phasing checking at high voltage should be tested immediately before and after use against a high voltage test supply.
3. Live voltage and phase checking on high voltage equipment may only be undertaken by a Authorised Person (HV), with assistance if necessary from a Competent Person (HV) acting on verbal instructions from the Authorised Person (HV). Neither a permit-to-work nor a sanction-for-test is required, but the Authorised Person (HV) and any assistant should be accompanied by an Accompanying Safety Person (HV).

Testing at high voltage:

1. Where high voltage tests are to be undertaken, a sanction-for-test should be issued to the Competent Person (HV) who is to be present throughout the duration of the tests.
2. The areas containing exposed live conductors, test equipment and any high voltage test connection should be regarded as high voltage enclosures.

High voltage test enclosures:

1. Unauthorised access to a high voltage test enclosure should be prevented by, as a minimum, red and white striped tape not less than 25 mm wide, suspended on posts, and by the display of high voltage danger signs. An Accompanying Safety Person (HV) or the Duty Authorised Person (HV) should be present throughout the duration of the tests, and the area should be continually watched while testing is in progress.

WORK ON BUSBAR SPOUTS OF MULTI-PANEL SWITCHBOARDS

When work is to be carried out on busbar spouts, the following operations should be carried out in strict sequence:

a. The Authorised Person (HV) should record the details of necessary safety precautions and switching operations on a safety programme and produce an isolation and earthing diagram;
b. The section of the busbar spouts on which work is to be carried out must be isolated from all points of supply from which it can be made live;
c. The isolating arrangements should be locked so that they cannot be operated, and shutters of live spouts locked shut. Caution signs should be fixed to the isolating points;
d. Where applicable, danger signs should be attached on or adjacent to the live electrical equipment at the limits of the zone in which work is to be carried out;
e. Busbars should be checked by means of an approved voltage indicator to verify that they are dead, the indicator itself being tested immediately before and after use. The checking with the voltage indicator should be done on the panel to which the circuit main earths are to be applied. This test should also be made on the panel on which the work is carried out;
f. Circuit main earths should be applied at a panel on the isolated section of the busbar other than that at which work is to be done using the method recommended by the switchgear manufacturers. The insertion of hands or any tool into the contact spouts for this purpose is not an acceptable practice;
g. An earth connection should also be applied to all phases at the point-of-work;
h. The permit-to-work should be issued to cover the work to be done. During the course of the work, where applicable, the earth connection(s) at the point-of-work may be removed one phase at a time. Each phase earth connection must be replaced before a second-phase earth connection is removed;
j. On completion of the work, the permit-to-work should be cancelled.

DEFINITION OF SAFETY TERMS:

Definitions presented here are those deemed necessary and suitable for electrical laboratory applications present in the Electronics and Electrical Engineering Laboratory. They should not be assumed to be directly related to definitions presented in other electrical standards or codes.

High Voltage: Any voltage exceeding 1000 V rms or 1000 V dc with current capability exceeding 2 mA ac or 3 mA dc, or for an impulse voltage generator having a stored energy in excess of 10 mJ. These current and energy levels are slightly below the startle response threshold.

Moderate Voltage: Any voltage exceeding 120 V rms (nominal power line voltage) or 120 V dc, but not exceeding 1000 V (rms or dc), with a current capability exceeding 2 mA ac or 3 mA dc.

Temporary Setups: Systems set up for measurements over a time period not exceeding three months.

Test Area: Area in which moderate voltages are accessible, and which has been clearly delineated by fences, ropes, and barriers.

Troubleshooting: Procedure during which energized bare connectors at moderate or high voltages might be temporarily exposed for the purpose of repair or problem diagnosis.

Inter lock: A safety circuit designed to prevent energizing high- or moderate-voltage power supplies until all access doors are closed, and to immediately de-energize such power supplies if the door is opened. Note that this function does not necessarily ensure full discharge of stored energy.

Bare Conductor: A conductor having no covering or electrical insulation whatsoever.

Covered Conductor: A conductor enclosed within a material of composition or thickness not defined as electrical insulation .

Insulated Conductor: A conductor encased within material of composition and thickness defined as electrical insulation.

Exposed Conductor: Capable of being inadvertently touched or approached nearer than a safe distance by a person. It applies to parts that are not suitably guarded, isolated, or insulated.

Unattended Operation: The operation of a permanent setup for electrical measurements for a time period longer than can be reasonably attended by staff.

Enclosed: Surrounded by a case, housing, fence or wall(s) that prevents persons from accidentally contacting energized parts.

TEMPORARY SETUPS

When troubleshooting a setup with exposed or bare conductors at high or moderate voltages, it may be necessary to temporarily bypass safety interlocks. Such procedures may only be performed under two-person operating conditions.

In instances where troubleshooting a system or particular equipment becomes frequent (at least once every six months) Group Leader approval is required. In all cases two staff members must be present when high voltage is energized and the interlock(s) bypassed. When troubleshooting a single piece of equipment in such a way that personnel may have access to high or moderate voltage (for example, repairing an instrument), two persons should be present. The "keep one hand in the pocket" rule is strongly encouraged.

SIGNS AND WARNING LIGHTS

DANGER HIGH VOLTAGE signs must be on display on all entrances to all test areas where bare conductors are present at both moderate and high voltages. These signs should be in the vicinity of the test area and on the outside of the door leading to the laboratory area. A warning light, preferably flashing, must be on when high or moderate voltages are present, and ideally should be activated by the energizing of the apparatus. The warning light must be clearly visible from the area surrounding the test area. In special cases where such a light interferes with an experiment, it can be omitted with special permission from the Group Leader and Division Chief.

In all cases where there is direct access from the outside hallway to the area where high or moderate voltages are present, a warning light, DANGER HIGH VOLTAGE sign, a safety interlock (for high voltages) and a locked door are required. For unattended setups with bare conductors at high or moderate voltage, a warning sign with the names of two contact persons and the dates of unattended operation must be posted on the door leading to the high-voltage area. In addition, written notice of unattended testing of high or moderate voltage with bare conductors must be sent to the NIST Fire Department (in Gaithersburg) or to the Engineering, Safety, and Support Division (in Boulder) clearly stating the anticipated dates of operation. A warning light on or near the door to the laboratory must be illuminated when high or moderate voltages with bare conductors are present.

GROUNDING STICK

Before touching a high-voltage circuit or before leaving it unattended and exposed, it must be de-energized and grounded with a grounding stick. The grounding stick must be left on the high-voltage terminal until the circuit is about to be re-energized.

Grounding sticks must be available near entrances to high-voltage areas. Automatic grounding arrangements or systems that employ audible warning tones to remind personnel to ground the high-voltage equipment are strongly encouraged for two-person operation, and are mandatory for one-person or unattended operation. For systems with bare conductors at moderate voltages, the use of a grounding stick is strongly recommended, particularly if the setup contains energy-storage devices.

MODES OF OPERATION

Two-person: Two-person operation is the normal mode of operation where high or moderated voltages are present. Allowed exceptions are: When all potentially dangerous voltages are confined inside a grounded or insulated box, or where the voltages are constrained in a shielded cable, or where the is no access to bare conductors

When one-person or unattended operation setups have been designed and approved according to the rules set out in this document and with appropriate approval. It is presumed that both individuals participating in two-person operation will follow basic high-voltage safety procedures and will monitor each other's actions to ensure safe behavior.

One-person: One-person operation of systems using high and moderate voltages with bare or exposed conductors, may be approved, after appropriate review and authorization, in order to provide for the efficient use of staff for long-term applications where it is judged that safety would not be compromised.

Unattended: It is recognized that in order to run efficient calibration services and maintain appropriate delivery schedules, unattended operation of systems using high and moderate voltages may be necessary. In such cases, unattended operation is permitted. with appropriate review and authorization, for systems having no bare or exposed conductors, and where required warning signs, lights, and barriers are present.

Unattended operation of setups with bare or exposed conductors at high and moderate voltages may be necessary under special circumstances, such as for unusually long data- acquisition periods. This is meant to be a rare occurrence. Should this mode of operation be frequently employed, then the apparatus should be modified to enclose all potentially dangerous voltages.

CIRCUIT BREAKERS & DISCONNECTS

Circuit breakers, disconnects or contactors used to energize a high-voltage source must be left in an open position when the supply is not in use. Laboratories should always be left in a configuration that at least two switches must be used to energize high-voltage circuits. Whenever possible a "return-to-zero-before energizing" interlock should be incorporated into the high-voltage supply.

Proper Circuit Design Recommendations

- Draw the circuit and study it before wiring it for operation at high voltage.
- Make sure all devices that require grounding are securely grounded.
- Allow adequate clearances between high-voltage terminals and ground.
- Solicit a second opinion before operation for the first time.

Transformers and Variacs:

- Make certain that one terminal of each transformer winding used to provide a separately derived system (this excludes the winding connected to the power supply) as well as the transformer or Variac case are properly grounded.
- The common terminal of a Variac should be connected to the supply neutral.
- Cascade transformers and, in some cases, isolation transformers are exceptions.

- General Information PERMIT-T0-WORK:– Issued by an authorised person to a responsible person who will perform the task of repair/ maintenance.
- Generally valid only for 24-Hrs. Permit to be re-validated by the permit-holder if work extends beyond 24 Hrs. after issue Formats will vary and be customized for a particular vessel/marine installation.

Permit To Work- BROAD GUIDELINES:

Prepared in duplicate copy and has at least five sections:

- 1st section states the nature of work to be carried out.
- 2nd section declares where electrical isolation and earthing have been applied and where Danger/Caution notices have been displayed.
- 3rd section is signed by the Person receiving the Permit acknowledging that he is satisfied with the safety precautions taken and the Isolation/ Earthing measures adopted.
- 4th section is signed by the Permit-holder that the work has been completed/suspended.
- 5th Section is signed by the Issuing authority cancelling the Permit.

HIGH VOLTAGE SAFETY AND PRECAUTIONS

For the purposes of safety, HV equipment includes the LV field system for a propulsion motor as it is an integrated part of the overall HV equipment. From the HV generators, the network supplies HV motors (for propulsion, side thrusters and air conditioning compressors) and the main transformer feeders to the 440 V switchboard. Further distribution links are made to interconnect with the emergency switchboard.

HV CIRCUIT BREAKERS AND CONTACTORS

Probably the main difference between a HV and an LV system occurs at the HV main switchboard. For HV, the circuit breaker types may be air-break, oil-break, gas-break using SF6 (sulphur hexafluoride) or vacuum-break. Of these types, the most popular and reliable are the vacuum interrupters, which may also be used as contactors in HV motor starters.

Each phase of a vacuum circuit breaker or contactor consists of a fixed and moving contact within a sealed, evacuated envelope of borosilicate glass. The moving contact is operated via flexible metal bellows by a charging motor/spring or solenoid operating mechanism. The high electric strength of a vacuum allows a very short contact separation, and a rapid restrike-free interruption of the arc is achieved.

When an alternating current is interrupted by the separating contacts, an arc is formed by a metal vapour from the material on the contact surfaces and this continues to flow until a current zero is approached in the a.c. wave form. At this instant the arc is replaced by a region of high dielectric strength which is capable of withstanding a high recovery voltage. Most of the metal vapour condenses back on to the contacts and is available for subsequent arcing. A small amount is deposited on the shield placed around the contacts which protects the insulation of the enclosure. As the arcing period is very short (typically about 15 ms), the arc energy is very much lower than that in air-break circuit-breakers so vacuum contacts suffer considerably less wear.

Because of its very short contact travel a vacuum interrupter has the following advantages:

- Compact quiet unit
- Minimum maintenance
- Non-flammable and non-toxic
- The life of the unit is governed by contact erosion but could be up to 20 years.

In the gas-type circuit breaker, the contacts are separated in an SF6 (sulphur hexafluoride) gas which is typically at a sealed pressure chamber at 500 kPa or 5 bar (when tested at 20° C).

HV INSULATION REQUIREMENTS

The HV winding arrangements for generators, transformers and motors are similar to those at LV except for the need for better insulating materials such as Micalastic or similar.

The HV windings for transformers are generally insulated with an epoxy resin/powdered quartz compound. This is a non-hazardous material which is maintenance free, humidity resistant and tropicalised. Conductor insulation for an HV cable requires a more complicated design than is necessary for an LV type. However, less copper area is required for HV conductors which allows a significant saving in space and weight for an easier cable installation. Where the insulation is air (e.g. between bare-metal live parts and earth within switchboards and in terminal boxes) greater clearance and creepage distances are necessary in HV equipment.

INSULATION RESISTANCE TESTS OF HV EQUIPMENT:

- A 5000 Vdc Megger, Hand-cranking or Electronic can be used for equipments upto 6.6KV.
- For routine testing of IR, 5000 Vdc must be applied for 1 minute either by cranking at constant speed with a Hand-cranking megger or by maintaining a 5000 Vdc continuously by a PB in an Electronic Megger.

IR values taken at different temperatures are unreliable, particularly if the temperature differences are more than 10°C.

SAFETIES OF IR TEST TO HV EQUIPMENTS

1. Before applying an IR test to HV equipment its power supply must be switched off, isolated, confirmed dead by an approved live-line tester and then earthed for complete safety.

2. The correct procedure is to connect the IR tester to the circuit under test with the safety earth connection ON. The safety earth may be applied through a switch connection at the supply circuit breaker or by a temporary earth connection local to the test point. This is to ensure that the operator never touches a unearthed conductor.
3. With the IR tester now connected, the safety earth is disconnected (using an insulated extension tool for the temporary earth). Now the IR test is applied and recorded. The safety earth is now reconnected before the IR tester is disconnected.

This safety routine must be applied for each separate IR test.

At prescribed intervals and particularly after a major repair work on an equipment or switchgear, a Polarisation Index(PI) may be taken to assess the condition of insulation of the equipment. PI readings are less sensitive to temperature changes.

POLARISATION INDEX (PI):

When the routine IR value tests (taken at different temperatures) are doubtful or during annual refit or after major repairs are undertaken, a PI test is conducted.

- PI value is the ratio between the IR value recorded after application of the test voltage continuously for 10 minutes to the value recorded after 1 minute of application.
- PI value= 2.0 or more is considered satisfactory.

A motor-driven megger is essential for carrying out a PI test.

HIGH VOLTAGE EQUIPMENT TESTING

The high voltage (e.g. 6.6 kV) installation covers the generation, main supply cables, switchgear, transformers, electric propulsion (if fitted) and a few large motors e.g. for side-thrusters and air conditioning compressors. For all electrical equipment the key indicator to its safety and general condition is its insulation resistance (IR) and this is particularly so for HV apparatus. The IR must be tested periodically between phases and between phases and earth. HV equipment that is well designed and maintained, operated within its power and temperature ratings should have a useful insulation life of 20 years. Large currents flowing through machine windings, cables, bus-bars and main circuit breaker contacts will cause a temperature rise due to I^2R resistive heating. Where overheating is suspected, e.g. at a bolted bus-bar joint in the main switchboard, the local continuity resistance may be measured and checked against the manufacturers recommendations or compared with similar equipment that is known to be satisfactory.

A normal ohmmeter is not suitable as it will only drive a few mA through the test circuit. A special low resistance tester or micro-ohmmeter (traditionally called a ducter) must be used which drives a calibrated current (usually I = 10

A) through the circuit while measuring the volt-drop (V) across the circuit. The meter calculates R from V/I and displays the test result. For a healthy bus-bar joint a continuity of a few mΩ would be expected.

Normally the safe testing of HV equipment requires that it is disconnected from its power supply. Unfortunately, it is very difficult, impossible and unsafe to closely observe the on-load operation of internal components within HV enclosures. This is partly resolved by temperature measurement with an recording infra-red camera from a safe distance. The camera is used to scan an area and the recorded infra-red image is then processed by a computer program to display hot-spots and a thermal profile across the equipment.

SAFETY TESTING OF HV EQUIPMENT:

Normally the safe testing of HV equipment requires that it is disconnected from its power supply. Unfortunately, it is very difficult, impossible and unsafe to closely observe the on-load operation of internal components within HV enclosures. This is partly resolved by temperature measurement with an recording infra-red camera from a safe distance. The camera is used to scan an area and the recorded infra-red image is then processed by a computer program to display hot-spots and a thermal profile across the equipment.

SANCTION-FOR-TEST SYSTEM

- Following work on a high voltage system, it is often necessary to perform various tests. testing should only be carried out after the circuit main earth (CME) has been removed.
- A sanction-for-test declaration should be issued in an identical manner to a permit to work provided and it should not be issued on any apparatus where a permit to work or where another sanction-fortest is in force.

Note That:

A sanction-for-test is not a permit to work.

An example of a sanction-for-test declaration is shown in the code of safe working practices (COSWP) 2010 edition annex 16.2.1.

ADDITIONAL PROCEDURES NEEDED FOR HV SYSTEMS

LIMITATION OF ACCESS FORM

When carrying out high voltage maintenance, it may be dangerous to allow anyone to work adjacent to high voltage equipment, as workers may not be familiar with the risks involved when working on or nearby high voltage equipment. The limitation of access form states the type of work that is allowed near high voltage equipment and safety precautions. the form is issued and signed by the chief engineer AND electrical officer, and countersigned by the persons carrying out the work.

Additional Procedures Needed for HV systems Earthing Down Earthing down is a very important concept to understand when working with high voltage systems. It is important to ensure that any stored electrical energy in equipment insulation after isolation is safely discharged to earth. the higher levels of insulation resistance required on high voltage cabling leads to higher values of insulation capacitance (c) and greater stored energy (w). this is demonstrated by the electrical formula: energy stored (w) joules = (capacitance x voltage^2)/ 2

Earthing down ensures that isolated equipment remains safe. Additional Procedures Needed for HV systems

There are two types of earthing down a high voltage switchboard:

1. CIRCUIT EARTHING
   - An incoming or outgoing feeder cable is connected by a heavy earth connection from earth to all three conductors after the circuit breaker has been racked out. This is done at the circuit breaker using a special key. This key is then locked in the key safe. The circuit breaker cannot be racked in until the circuit earth has been removed.
2. BUSBAR EARTHING
   - When it is necessary to work on a section of the busbars, they must be completely isolated from all possible electrical sources. This will include generator incoming cables, section or bus-tie breakers, and transformers on that busbar section. The busbars are connected together and earthed down using portable leads, which give visible confirmation of the earthing arrangement.

High voltage safety checklists for the following can be found in onboard "Company Safety Manual" and sample can be found in the "Code of Safe working Practices for Merchant Seaman (COSWP)" 2010 edition:

- Working on high voltage equipment/installations
- Switchgear operation
- Withdrawn apparatus not being used
- Locking off
- Insulation testing
- Supply failure
- Entry to high voltage enclosures
- Earthing
- Working on high voltage cables
- Working on transformers
- Safety signs
- Correct personal protective equipment

Personnel should not work on High Voltage equipment unless it is dead, isolated and earthed at all high voltage disconnection points. The area should be secured, permits to work or sanction for test notices issued, access should be limited and only competent personnel should witness the testing to prove isolation.

WORK PROCEDURES IN HIGH VOLTAGE

Working procedures are divided in to three distinct groups.

1. Dead working
2. Live working
3. Working in the vicinity of live parts

Dead Working:

Work activity on electrical installations which are neither live nor charged, carried out after taking all measures to prevent electrical danger.

Precautions before starting work

- Obtain PTW/Sanction- to-Test Permit before commencing work
- Test and prove that the equipment is DEAD before earthing. (with a HV line tester)
- Earth the equipment

Working in the vicinity of live parts:

- All work activity in which the worker enters the vicinity of live zone with his body or with tools and equipment without encroaching in to live zone.
- Using the correct personal protective equipment (PPE) and following safe work practices will minimize risk of electrical shock hazards

HIGH VOLTAGE EQUIPMENT

A typical high voltage installation will incorporate only high voltage rated equipment on the following:

1. Generating sets
2. High voltage switchboards with associated switchgear, protection devices and instrumentation high voltage cables
3. High voltage/low voltage step-down transformers to service low voltage consumers
4. High voltage/high voltage (typically 6.6kV/2.9kV) step-down transformers supplying propulsion converters and motors
5. High voltage motors for propulsion, thrusters, air conditioning and compressors

A high voltage electrical shock is a significant danger to any person carrying out electrical work. Any simultaneous contact with a part of the body and a live conductor will probably result in a fatal electric shock. There is also a risk of severe burn injuries from arcing if conductors are accidentally short-circuited. A high voltage electric shock will almost certainly lead to severe injury or a fatality.

Factors that could increase the risk of receiving an electric shock:

1. High voltage work may be carried out close to a person that is not familiar with high voltage hazards. therefore, the area must be secured from the surrounding non-electrical work and danger notices posted.
2. Areas of earthed metal that can be easily touched increase the possibility of electric shock from a high voltage conductor.

DANGERS WORKING WITH HIGH VOLTAGE EQUIPMENTS

High voltage insulation testing (flash testing) can be particularly hazardous when several parts of the equipment are energised for a period of time. Equipment using water as part of the high voltage plant can lead to an increased risk of injury.

Using test instruments when taking high voltage measurements can increase the risk of injury if the protective earth conductor is not connected. This can result in the enclosure of the instrument becoming live at dangerous voltages. High voltage equipment will store energy after disconnection. for example, on a 6.6kv switchboard, a fatal residual capacitive charge may still be present hours or even days later. If, during maintenance, a high voltage circuit main earth is removed from the system, it must not be worked on as the high voltage cabling can recharge itself to a high voltage (3–5kv).

DANGERS WORKING WITH HIGH VOLTAGE EQUIPMENTS

TRANSFORMER TESTING & MAINTENANCE

Transformer is a static device which transforms a.c. electrical power from one voltage to another voltage keeping the frequency same by electromagnetic induction.

Main features of transformer:

Outdoor,oil cooled, 3 phase,50hz

Primary is delta connected and secondary is star connected.

Naturaly cooled

Amongst all the types of transformers this is the most required and most used type.

Parts of transformer:

- Main tank
- Radiators
- Conservator
- Explosion vent
- Lifting lugs
- Air release plug
- Oil level indicator
- Tap changer

- Wheels
- Hv/lv bushings
- Filter valves
- Oil filling plug
- Drain plug
- Cable box

TESTING OF TRANSFORMER:

- Testing is carried out as per PMS or Company checklist.
- Routine, type tests & special tests
- Routine tests (to be carried out on each job):

1. Measurement of winding resistance
2. Measurement of insulation resistance
3. Seperate source voltage withstand test (high voltage tests on HV & LV)
4. Induced over voltage withstand test (dvdf test)
5. Measurement of voltage ratio
6. Measurement of no load loss & current.
7. Measurement of load loss & impedence.(efficiency & regulation)
8. Vector group verification
9. Oil bdv test.
10. Tests on oltc (if attached)

MAINTENANCE OF TRANSFORMER

- Transformer is the heart of any power system. Hence preventive maintenance is always cost effective and time saving. Any failure to the transformer can extremely affect the whole functioning of the organization.

MAINTENANCE PROCEDURE

OIL:

1. Oil level checking. Leakages to be attended.
2. Oil BDV & acidity checking at regular intervals. If acidity is between 0.5 to 1mg KOH, oil should be kept under observation.
3. BDV, Color and smell of oil are indicative.
4. Sludge, dust, dirt,moisture can be removed by filtration.
5. Oil when topped up shall be of the same make. It may lead to sludge formation and acidic contents.
6. Insulation resistance of the transformer should be checked once in 6 months.
7. Megger values along with oil values indicate the condition of transformer.
8. Periodic Dissolved Gas Analysis can be carried out.

MAINTENANCE BUSHINGS

1. Bushings should be cleaned and inspected for any cracks.
2. Dust & dirt deposition, Salt or chemical deposition, cement or acid fumes depositions should be carefully noted and rectified.

MAINTENANCE

1. Periodic checking of any loose connections of the terminations of HV & LV side.
2. Breather examination. Dehydration of Silica gel if necessary.
3. Explosion vent diaphragm examination.
4. Conservator to be cleaned from inside after every three years.
5. Regular inspection of OIL & WINDING TEMPERATURE METER readings.
6. Cleanliness in the Substation yard with all nets, vines, shrubs removed.

WORK ON DISTRIBUTION TRANSFORMERS

When work is to be carried out on the connections to, or the windings of, a distribution transformer:

a. The Authorised Person (HV) should record the details of necessary safety precautions and switching operations on a safety programme, and produce an isolation and earthing diagram;
b. The switchgear or fuse gear controlling the high voltage windings should be switched off, and a safety lock and caution sign fitted;
c. The low voltage windings of the transformer switch or isolator should be switched off, and a safety lock and caution sign fitted, or other physical means should be used to prevent the switch being energised during the course of work;
d. Where applicable, danger signs should be attached on or adjacent to the live electrical equipment at the limits of the zone in which work is to be carried out;
e. The transformer should be proved dead at the points-of-isolation if practicable;
f. An earth should then be applied to the high voltage winding via the switchgear and a safety lock fitted. If the proprietary earthing gear is available for the low voltage switchgear, it should be fitted and safety locks applied (it is advisable to retest for dead before fitting this earthing gear);
g. Before a permit-to-work is issued – the Authorised Person (HV) should, at the point- of-work in the presence of the Competent

Person (HV), identify and mark the transformer to be worked on. The permit-to-work and the key to the key safe should then be issued to the Competent Person (HV);

PROTECTION OF TRANSFORMERS

1. The best way of protecting a transformer is to have good preventive maintenance schedule.
2. Oil Temperature Indicators.
3. Winding Temperature indicators.
4. Buchholz Relay.
5. Magnetic Oil level Gauge.
6. Explosion Vent.
7. HT fuse & D.O. fuse.
8. LT circuit breaker.
9. HT Circuit breaker with Over load, Earth Fault relay tripping.
10. Oil Surge Relay for OLTC.
11. PRV for OLTC.
12. HORN GAPS & Lightening Arrestor.
13. Breather.

FAILURES & CAUSES

- Insufficient Oil level.
- Seepage of water in oil.
- Prolonged Over loading.
- Single Phase loading.
- Unbalanced loading.
- Faulty Termination (Improper sized lugs etc)
- Power Theft.
- Prolonged Short Circuit.
- Faulty operation of tap changer switch.
- Lack of installation checks.
- Faulty design
- Poor Workmanship
  - Improper formation of core.
  - Improper core bolt insulation.
  - Burr to the lamination blades.
  - Improper brazing of joints.
  - Burr/sharp edges to the winding conductor.
  - Incomplete drying.
  - Bad insulation covering.
  - Insufficient cooling ducts in the winding.
- Bad Quality of raw material.
- Transit damaged transformers.
- After failure, transformer is removed and replaced with new/repaired one without removing the cause of failure which results in immediate or short time failure.

HIGH VOLTAGE EQUIPMENT MAINTENANCE

MAINTENANCE OF SWITCHGEAR ENCLOSURES

Strictly adhere to required procedures for system switching operations. Switching, de-energizing and energizing shall be performed by authorized personnel only. Install temporary grounding leads for safety. Remove necessary access and coverplates. Fill out inspection test form. Record data in reference to equipment. Completely isolate switchgear enclosure to be tested and inspected from sources of power.

Mechanical Inspection:

- Check mechanical operation of devices.
- Check physical appearance of doors, devices, equipment and lubricate in accordance with manufacturer's instructions.
- Check condition of contacts.
- Check disconnects, starters, and circuit breakers in accordance with inspection and test reports and procedures.
- Check condition of bussing for signs of overheating, moisture or other contamination, for proper torque, and for clearance to ground.
- Inspect insulators and insulating surfaces for cleanliness, cracks, chips, tracking.
- Report discovered unsafe conditions.
- Remove drawout breakers and check drawout equipment.
- Check cable and wiring condition, appearance, and terminations. Perform electrical tests as required.
- Inspect for proper grounding of equipment.
- Perform breaker and switch inspection and tests

Cleaning:

i. Check for accumulations of dirt especially on insulating surfaces and clean interiors of compartments thoroughly using a vacuum or blower.
ii. Remove filings caused by burnishing of contacts.
iii. Do not file contacts. Minor pitting or discoloration is acceptable.
iv. Report evidence of severe arcing or burning of contacts.
v. Degrease contacts with suitable cleaners

Electrical Testing:

i. Check electrical operation of pilot devices, switches, meters, relays, auxiliary contacts, annunciator devices, flags, interlocks, cell switches, cubicle lighting. Visually inspect arrestors, C/T's and P/T's for signs of damage. Record data on test report form.
ii. Megger test insulators to ground.
iii. Megger test bussing phase to ground, and phase to phase, using a 1000 volt megger.
iv. DC hipot phases to others and to ground using step voltage method

as specified for cables with withstand levels held for not less than one minute. Record decay curve, current versus time to completion of test, and indicate withstand level.

Electrical Testing:v. Maximum DC hipot test levels shall be as follows:

a) 25kV class 50kV DC
b) 15kV class 28.5kV DC
c) 5kV class 9kV DC
d) Test contact resistance across bolted sections of buss bars. Record results and compare test values to previous acceptance and maintenance results and comment on trends observed.

At completion of inspection and test, remove temporary grounds, restore equipment to serviceable condition and recommission equipment. Compare test results to previous maintenance test results.

MAINTENANCE OF HIGH VOLTAGE AIR/OIL CIRCUIT BREAKERS:

1. Strictly adhere to required procedures for system switching operations. Switching, de-energizing and energizing shall be performed by authorized personnel only.

Completely isolate circuit breakers to be worked on from power sources.

Install temporary grounds.

Remove circuit breaker from cubicle unless bolted type.

Record manufacturer, serial number, type and function of breaker, reading of operations counter, date of inspection, and signature of person responsible for inspection on report sheet.

Mechanical Inspection:

Inspect for:

- Accumulations of dirt, especially on insulating surfaces.
- Condition of primary contact clusters.
- Condition of control wiring plug-in contacts.
- Condition of moving and fixed main contacts, excessive heating or arcing.
- Condition of arcing contacts.
- Cracks or indications of tracking on insulators.
- Tracking or mechanical damage to interphase barriers.
- Flaking or chipping of arc chutes.
- Broken, damaged or missing springs on operating mechanism.
- Damage to or excessive wear on operating linkage, ensure all clevis pins are securely retained in position.

Inspect for:

- Correct alignment of operating mechanism and contacts.
- Evidence of corrosion and rusting of metals, and deterioration of painted surfaces.

- Oil breakers only:
  a) Refer to manufacturer's maintenance manual for special tools that may be required to check oil breaker contacts.
  b) Check oil holding tanks in accordance with manufacturer's instructions.
  c) Check for proper oil level and condition of level gauge.

Cleaning:

i. Remove accumulations of dirt from insides of cubicles with vacuum cleaner and/or blower.
ii. Clean insulating surfaces using brush or wiping with lint free cloth.8. Check fixing bolts of hardware and breaker components for tightness.

'Dress' pitting on contact surfaces, using a burnishing tool. 'Dress' major arcing on contacts to smooth condition. Remove filings before switchgear is re-energized. Report unsafe conditions resulting from severe arcing or burning of contacts. On completion of foregoing tasks, lightly lubricate bearing points in operating linkage with manufacturer's specified lubricant. Operate breaker several times to ensure smoothness of mechanical operation.11. Check potential and current transformer cable connections for tightness.

Replace inspection lamp where fitted. On first inspection, record data to auxiliary equipment, i.e. primary fuses, potential transformer, potential fuses, and current transformers. Record serial numbers, catalogue numbers, sizes, ratios. On completion of inspection and test, remove temporary grounds. Restore equipment to serviceable condition.

Electrical Maintenance Tests:

a) *General:*
   i. Test contact resistance across closed line-load contacts, and line and load circuit breaker plug-in clusters. Record results. Clean contacts using appropriate tools to get lowest contact resistance reading possible.
   ii. Test insulation resistance for all phases to others and to ground.
   iii. Test electrical function in accordance with breaker manufacturer's instructions and drawings.
b) *Air Breakers:*
   i. Prior to hipot test being carried out, ensure surrounding primary connections to main equipment are properly grounded and isolated.
   ii. DC hipot test at test levels indicated for switchgear enclosure.
c) *Oil Breakers:*
   i. Do not perform DC hipot tests on oil circuit breakers.
   ii. Dielectric (hipot) test on insulating oil per ASTM D877. Compare dielectric strength test results to previous test data where applicable, and comment on changes.

FUSED OR UNFUSED LB AND NLB DISCONNECT SWITCHGEAR

1 Strictly adhere to required procedures for system switching operations. Switching, de-energizing and energizing shall be performed by authorized personnel only.
2 Completely isolate switchgear to be worked on from power sources.
3 Remove access covers and plates.
4 Test and discharge equipment to be worked on.
5 Install temporary safety grounds.
6 Report manufacturer, serial number, type, function of switchgear assembly, date of inspection, and signature of person responsible for inspection.

Mechanical Inspection: inspect for:

- Accumulations of dirt, especially on insulating surfaces.
- Condition of moving and fixed contact, excessive heating or arcing.
- Cracks, or tracking on insulators.
- Tracking or mechanical damage to interphase barriers.
- Chipping or flaking of arc chutes or arc shields.
- Fixing bolts being fully tightened where bolted-on shields are fitted.
- Overheating or arcing on fuses and fuse holders.
- Correct fuse clip tension.
- Broken, missing or damaged springs on operating mechanism.
- Damage to or excessive wear on operating linkage. Check that all clevis pins are securely retained in position.
- Correct alignment of contact blades and operating linkage.
- Corrosion & rusting of metals, deterioration of painted surfaces.
- Proper operation of key interlock or other mechanical interlock (if applicable).
- Evidence of corona deterioration.

Cleaning:

- Remove accumulations of dirt from insides of switchgear cubicles using vacuum cleaner and/or blower.
- Clean insulating surfaces using brush or wiping with lint free cloth.
- Do not file contacts. Minor pitting or discoloration is acceptable.
- Report evidence of severe arcing or burning of contacts.
- Degrease contacts with suitable cleaners.

Check that connections, including current limiting fuses, are secure. Torque to manufacturer's requirement.

Electrical Maintenance Tests:

- Test insulation resistance for all phases to others and to ground.
- Test contact resistance across switch blade contact surfaces.
- Test electrical charging mechanism of switch if applicable.
- Test electrical interlocks for proper function.

- DC hipot test phases to the others and to ground using step method to levels specified for switchgear.
- Operate blown fuse trip devices if applicable.11. After testing is completed, remove temporary grounds and restore equipment to serviceable condition.

MAINTENANCE OF PROTECTION RELAYS

1. Strictly adhere to required procedures for system switching operations. Switching, de-energizing and energizing shall be performed by authorized personnel only.
2. Completely isolate protective relays to be tested and inspected from sources of power.
3. Set and test protective relays to “as found” settings or to new settings provided by Minister prior to maintenance commissioning.
4. Use manufacturer’s instructions for information concerning connections, adjustments, repairs, timing, and data for specific relay.

Mechanical Inspection of Induction Disc Relays:

- Carefully remove cover from relay case. Inspect cover gasket. Check glass for tightness and cracks.
- Short-circuit current transformer secondary by careful removal of relay test plug or operation of appropriate current blocks.
- Ensure disc has proper clearance and freedom of movement between magnet poles.
- Check connections and taps for tightness.
- Manually operate disc to check for freedom of movement. Allow spring to return disc to check proper operation.
- Check mechanical operation of targets.
- Check relay coils for signs of overheating and brittle insulation

Cleaning:

- Clean glass inside and out.
- Clean relay compartment as required. Clean relay plug in contacts, if applicable, using proper tools.
- Remove dust and foreign materials from interior of relay using small brush or low pressure (7 lbs.) blower of nitrogen.
- Remove rust or metal particles from disc or magnet poles with magnet cleaner or brush.
- Inspect for signs of carbon, moisture and corrosion.
- Clean pitted or burned relay contacts with burnishing tool or non-residue contact cleaner.

Electrical Testing: Tests for typical overcurrent relays include:

- Zero check.
- Induction disc pickup.
- Time-current characteristics.

- Target and seal-in operation.
- Instantaneous pickup.
- Check C/T & P/T ratios and compare to coordination data.
- Proof test each relay in its control circuit by simulated trip tests to ensure total and proper operation of breaker and relay trip circuit by injection of the relay circuit to test the trip operation.

Solid State Relays:

- Inspect and test in accordance with manufacturer's most recent installation and maintenance brochure.
- Perform tests using manufacturer's relay test unit as applicable, with corresponding test instructions.
- If the manufacturer's tester is not available, use a relay tester unit approved by relay manufacturer, with proper test data and test accessories.
- Proof test each relay in its control circuit by simulated trip tests to ensure total and proper operation of breaker and relay trip circuit by injection of relay circuit to test trip operation.
- Check C/T and P/T ratios and compare to coordination date.9. At completion of inspection and test, restore equipment to serviceable condition and recommission equipment. Compare test results to previous maintenance test results.

MAINTENANCE OF OVERHEAD RADIAL POWER LINES:

1. Strictly adhere to required procedures for system switching operations. Switching, de-energizing and energizing shall be performed by authorized personnel only.
2. Completely isolate overhead radial power lines to be tested and inspected from sources of power.
3. Install temporary grounding leads for safety.
4. Inspect insulators and insulating surfaces for cleanliness, cracks, chips, tracking, and clean insulators thoroughly.
5. Check cable connections to insulators and check cable sag between poles. Report discovered unsafe conditions.
6. Visually check wooden poles and sound test with 18 oz. wooden mallet.
7. Visually inspect metal line structures for rust, deterioration, metal fatigue, and report discovered unsafe conditions.
8. Inspect crossarms, bolts, rack assemblies, guys, guy wires, and dead ends. Report discovered unsafe conditions.
9. Visually inspect grounding connections.
10. On completion of inspection, remove temporary grounding, restore equipment to serviceable condition

SURGE ARRESTORS:

1. Strictly adhere to required procedures for system switching

operations. Switching, de-energizing and energizing shall be performed by authorized personnel only.

2. Completely isolate surge arrestors to be tested and inspected from sources of power.
3. Install temporary grounding leads for safety.
4. Inspect surge arrestors for cleanliness, cracks, chips, tracking and clean thoroughly.
5. Perform insulation power factor test. Record results.
6. Perform grounding continuity test to ground grid system, record results.
7. On completion of inspection and testing, remove temporary grounds, restore equipment to serviceable condition.

DISCONNECTION PROCEDURE:

Safety of Disconnection Switch:

1. When a disconnect switch is installed in this manner, the frame of the disconnect switch, the upper and lower steel operating rod and the switch handle are all bonded together and connected to the common neutral and the pole's ground rod, effectively eliminating any insulating value of the insulated insert. The electrical worker operating the switch has no protection and could have as much as full system voltage from the worker's hands on the switch handle to the worker's feet.
2. The use of rated rubber gloves can eliminate touch potential if the switch were to fail and go to ground. But there is also the hazard of step potential for the worker operating the switch, and rated rubber gloves does nothing to eliminate step potential. Also, the maximum ASTM rating for rubber gloves is limited to 36 kV, eliminating worker protection from higher voltages.
3. Properly installed ground mats provide the best protection for workers operating disconnect switches while standing on the ground.

If the disconnect switch were to fail and go to ground, the switch handle could be energized at potentially full system voltage, say 7,200 volts, energizing the switch handle at 7,200 volts, less the voltage drop in the grounding conductor from the switch handle to the ground mat (typically 20 to 25 volts).

- But if the worker were wearing rated rubber gloves and standing on a ground mat attached to the switch handle, would they be safe? Yes!
- If they were not wearing rated rubber gloves but still standing on a ground mat attached to the switch handle, would they be safe? Yes!
- When the worker wears rated rubber gloves while standing on a ground mat attached to the switch handle, the gloves are insulating the worker from the 20 to 25 volts developed across the ground mat

and switch handle; well below any hazardous voltage. They are safe with or without rated rubber gloves if they are standing on a ground mat properly connected to the switch handle.

HV DISCONNECTION PROCEDURE:

Almost every major line or equipment in a substation has associated with it a means of completely isolating it from other energized elements as a prudent means of insuring safety by preventing accidental energization. These simple switches, called disconnects, or disconnecting switches. They are usually installed on both sides of the equipment or line upon which work is to be done.

How to operate these switches:

1. They should not be operated while the circuit in which they are connected is energized, but only after the circuit is deenergized.
2. They may be opened by means of an insulated stick that helps the operator keep a distance from the switch.
3. Locking devices are sometimes provided to keep the disconnects from being opened accidentally or from being blown open during periods of heavy fault currents passing through them.

ISOLATION OF ANY HIGH VOLTAGE EQUIPMENT:

What is isolation:

Isolation is a means of physically and electrically separating two parts of a measurement device, and can be categorized into electrical and safety isolation. Electrical isolation pertains to eliminating ground paths between two electrical systems. By providing electrical isolation, you can break ground loops, increase the common-mode range of the data acquisition system, and level shift the signal ground reference to a single system ground. Safety isolation references standards have specific requirements for isolating humans from contact with hazardous voltages. It also characterizes the ability of an electrical system to prevent high voltages and transient voltages from transmitting across its boundary to other electrical systems with which you can come in contact.

1. Isolation of individual circuits protected by circuit breakersWhere circuit breakers are used the relevant device should be locked-off using an appropriate locking-off clip with a padlock which can only be opened by a unique key or combination. The key or combination should be retained by the person carrying out the work.

Some DBs are manufactured with 'Slider Switches' to disconnect the circuit from the live side of the circuit breaker. These devices should not be relied upon as the only means of isolation for circuits as the wrong switch could easily be operated on completion of the work.

Isolation of individual circuits protected by fuses Where fuses are used, the simple removal of the fuse is an acceptable means of disconnection. Where removal of the fuse exposes live terminals that can be touched, the incoming supply to the fuse will need to be isolated. To prevent the fuse being replaced by others, the fuse should be retained by the person carrying out the work, and a lockable fuse insert with a padlock should be fitted as above. A caution notice should also be used to deter inadvertent replacement of a spare fuse. In addition, it is recommended that the enclosure is locked to prevent access as stated above under 'Isolation using a main switch or distribution board (DB) switch-disconnector'.Note

In TT systems, the incoming neutral conductor cannot reliably be regarded as being at earth potential. This means that for TT supplies, a multi-pole switching device which disconnects the phase and neutral conductors must be used as the means of isolation. For similar reasons, in IT systems all poles of the supply must be disconnected. Single pole isolation in these circumstances is not acceptable.

High voltage insulation testing (flash testing) can be particularly hazardous when several parts of the equipment are energized for a period of time.

Isolation Procedure:

1. Isolate from all sources of supply.
2. Prevent unauthorised connection by fixing safety locks and caution signs at points-of- isolation.
3. Fix danger signs on live equipment adjacent to the point-of-work.

PROVING THE SYSTEM IS DEAD:

How to prove:

Before starting work it should be proved that the parts to be worked on and those nearby are dead. It should never be assumed that equipment is dead because a particular isolation device has been placed in the off position.

1. The procedure for proving dead should be by use of a proprietary test lamp or two pole voltage detectors.
2. Non-contact voltage indicators (voltage sticks) and multi-meters should not be used.
3. The test instrument should be proved to be working on a known live source or proprietary proving unit before and after use.
4. All phases of the supply and the neutral should be tested and proved dead.

Proving dead unused or unidentified cables

- Where there is uncertainty regarding isolation when removing unidentified cables or proving dead an 'unused' cable, particularly where insufficient conductor is exposed to enable the use of test probes, those conductors should be assumed to be live until positively proven to be dead and any work carried out on them should employ live working practices until the conductors are proved dead.
- Clamp meters can be used as a means of identifying cables by testing for current flow in the conductors.

- Non-contact voltage indicators (voltage sticks) can also be useful in these situations to test for voltage where cables without a metallic sheath are to be identified. However, once insulation is pared using live working practices to reveal the underlying conductors, contact voltage detectors should be used as the means of proving dead.

Prove Dead:

(i) Prove dead with a high voltage potential indicator at all accessible points-of-isolation.

(ii) Where appropriate, prove dead on the low voltage side of a transformer, that is LV feed pillars, LV distribution boards etc.

EARTHING AND DISCHARGING OF HV:

Earthing down is a very important concept to understand when working with high voltage systems. It is important to ensure that any stored electrical energy in equipment insulation after isolation is safely discharged to earth. The application of earthing on high voltage conductors is controlled in accordance with the provisions of the Power System Safety Rules. The following general requirements and principles are applicable for portable earthing.

Safety:

1. Always carry earthing equipment below shoulder level;
2. Ensure that clamps and leads are kept a safe distance from any high voltage conductor;
3. Apply clamps to stirrup (if provided) or a horizontal conductor where possible;
4. Avoid clamp application to bushing caps and to braids; and
5. Position clamp so that tension on the earth lead is minimised.
6. Assemble and inspect earthing equipment on the ground;
7. Extend or otherwise prepare any earthing equipment such as shot gun sticks on the ground.
8. Proving High Voltage Conductors De-Energised
9. Do not allow any part of the earthing system to encroach on safe approach distances. Where practicable, keep the earthing leads away from the body;
10. Earthing equipment is to be removed carefully from high voltage conductors to prevent the equipment encroaching on or coming into contact with adjacent live high voltage conductors.
11. Check that the rating of the earthing equipment is appropriate for the fault level at the location at which it is to be applied.
12. Ensure that the earthing equipment is in a serviceable condition. Any portable earthing equipment found to be defective shall be removed from service for repair or disposal.

There are two types of earthing down a high voltage switchboard:

Circuit Earthing

- An incoming or outgoing feeder cable is connected by a heavy earth connection from earth to all three conductors after the circuit breaker has been racked out. This is done at the circuit breaker using a special key. This key is then locked in the key safe. The circuit breaker cannot be racked in until the circuit earth has been removed.

Busbar Earthing

- When it is necessary to work on a section of the busbars, they must be completely isolated from all possible electrical sources. This will include generator incoming cables, section or bus-tie breakers, and transformers on that busbar section. The busbars are connected together and earthed down using portable leads, which give visible confirmation of the earthing arrangement.

Earthing Procedure:

(i) Earth conductors at all points-of-isolation and fix safety locks to earths.
(ii) Identify with certainty or spike underground cables at the point/s of work if the conductors are to be cut or exposed.
(iii) Earth overhead lines near the working places.

ISSUING OF A PERMIT-TO-WORK OR SANCTION- FOR-TEST:

1. Before a permit-to-work or a sanction-for-test is issued, the Authorised Person or Electrical Engineer should identify the equipment on which the work or test is to be undertaken.
2. If the work involves, or may involve, obtaining access to items of equipment over which confusion could occur, the Authorised Person (HV)/ Electrical Engineer should identify such items to the Competent Person (HV) and apply temporary marking to them.
3. Before issuing a permit-to-work or sanction-for- test, the Authorised Person (HV)/ Electrical Engineer should show the Competent Person (HV) the isolation and earthing diagram and indicate the safety arrangements at the points-of-isolation and at the point-of-work or test.
4. The Authorised Person (HV)/ Electrical Engineer should ensure that the Competent Person (HV) understands all the relevant safety procedures and precautions.
5. If the Competent Person (HV) thereafter accepts the permit or sanction, that person becomes responsible for the defined work or test until the permit or sanction is cancelled.
6. Mark the point-of-work.

7. Issue the permit-to-work, isolation and earthing diagram, and the key to the safety key box to the Competent Person (HV).
8. Authorised Persons (HV)/ Electrical Engineer undertaking tasks requiring permits-to-work or sanctions-for-test should issue the documents to themselves.
9. Adjust mimic diagram and complete the site logbook.
10. All such documents must be countersigned by a site- certified Authorised Person (HV)/ Electrical Engineer before the work or test starts.

UNDERTAKE THE WORK:

The Competent Person (HV) is to undertake or directly supervise the work and on completion, or when the work is stopped and made safe, is to return the original of the permit-to-work, the isolation and earthing diagram and the Competent Person's (HV) key to the safety key box to the Duty Authorised Person (HV)/ Electrical Engineer, and complete part 3 of the permit retained in the pad.

CHECK THE EQUIPMENT:

If the work has been completed, check to ensure it is safe to energize. If the work has been stopped, check the equipment has been made safe.

CANCEL THE PERMIT-TO-WORK:

(i) Cancel the permit-to-work by signing the "completion of work" part and by cancel the permit in the presence of the Competent Person (HV).
(ii) File the isolation and earthing diagram in the operational procedure manual and permit-to-work in respective file.
(iii) Return key to key safe.

ISSUE THE SANCTION-FOR-TEST:

(i) The Competent Person (HV) is to be shown the safety arrangements at all the point/s of isolation and at the locations of the test, and is to initial the isolation and earthing diagram.
(ii) Issue the sanction-for-test, isolation and earthing diagram, and the key to the safety key box to the Competent Person (HV).
(iii) Retain working lock keys, and remove and replace earths as requested.

CANCEL THE SANCTION-FOR-TEST:

(i) Cancel the sanction-for-test by signing part 4 and by destroying the sanction in the presence of the Competent Person (HV).

(ii) File the isolation and earthing diagram in the operational procedure manual.
(iii) Return key to key safe.

RE-ENERGIZING EQUIPMENT:

i. Conduct tests and visual inspections to ensure all tools, electrical jumpers, shorts, grounds, and other such devices have been removed.
ii. Warn others to stay clear of circuits and equipment.
iii. Each lock and tag must be removed by the person who applied it.
iv. Visually check that all employees are clear of the circuits and equipment.

PROTECTIONS OPERATING PRINCIPLES

SF6 Circuit Breaker:

A circuit breaker in which the current carrying contacts operate in sulphur hexafluoride or SF6 gas is known as an SF6 circuit breaker. SF6 has excellent insulating property. SF6 has high electro-negativity. That means it has high affinity of absorbing free electron. Whenever a free electron collides with the SF6 gas molecule, it is absorbed by that gas molecule and forms a negative ion.

The attachment of electron with SF_{6gas} molecules may occur in two different ways, These negative ions obviously much heavier than a free electron and therefore over all mobility of the charged particle in the SF_6 gas is much less as compared other common gases. We know that mobility of charged particle is majorly responsible for conducting current through a gas.

Working of SF_6 Circuit Breaker:

The working of SF_6 CB of first generation was quite simple it is some extent similar to air blast circuit breaker. Here SF_6 gas was compressed and stored in a high pressure reservoir. During operation of SF_6 circuit breaker this highly compressed gas is released through the arc in breaker and collected to relatively low pressure reservoir and then it pumped back to the high pressure reservoir for re utilize. The working of SF_6 circuit breaker is little bit different in modern time. Innovation of puffer type design makes operation of SF_6 CB much easier. In buffer type design, the arc energy is utilized to develop pressure in the arcing chamber for arc quenching.

During opening of the breaker the cylinder moves downwards against position of the fixed piston hence the volume inside the cylinder is reduced which produces compressed SF_6 gas inside the cylinder. The cylinder has numbers of side vents which were blocked by upper fixed contact body during closed position. As the cylinder move further downwards, these vent openings cross the upper fixed contact, and become unblocked and then compressed SF_6

gas inside the cylinder will come out through this vents in high speed towards the arc and passes through the axial hole of the both fixed contacts. The arc is quenched during this flow of SF_6 gas.

During closing of the circuit breaker, the sliding cylinder moves upwards and as the position of piston remains at fixed height, the volume of the cylinder increases which introduces low pressure inside the cylinder compared to the surrounding. Due to this pressure difference SF_6 gas from surrounding will try to enter in the cylinder. The higher pressure gas will come through the axial hole of both fixed contact and enters into cylinder via vent and during this flow; the gas will quench the arc.

PROTECTIONS OPERATING PRINCIPLES

Principles of Distance Relays

Since the impedance of a transmission line is proportional to its length, for distance measurement it is appropriate to use a relay capable of measuring the impedance of a line up to a predetermined point (the reach point). Such a relay is described as a distance relay and is designed to operate only for faults occurring between the relay location and the selected reach point, thus giving discrimination for faults that may occur in different line sections.

The basic principle of distance protection involves the division of the voltage at the relaying point by the measured current. The apparent impedance so calculated is compared with the reach point impedance. If the measured impedance is less than the reach point impedance, it is assumed that a fault exists on the line between the relay and the reach point. The reach point of a relay is the point along the line impedance locus that is intersected by the boundary characteristic of the relay.

Since this is dependent on the ratio of voltage and current and the phase angle between them, it may be plotted on an R/X diagram. The loci of power system impedances as seen by the relay during faults, power swings and load variations may be plotted on the same diagram and in this manner the performance of the relay in the presence of system faults and disturbances may be studied.

Solid State Switching Principle

- High voltage testing does not usually require high power.
- Thus special methods may be used which are not applicable.
- Then generating high voltage in high power applications.
- In the field of electrical eng. & applied physics, high voltages are required for several applications as:

1. A power supply (eg. hv dc) for the equipments such as electron microscope and x-ray machine.
2. Required for testing power apparatus – insulation testing.

- High impulse voltages are required for testing purposes to simulate over voltages due to lightning and switching.

Solid State Switching Principle

Solid State Switching Principle

- Both full-wave as well as half-wave circuits can produce a maximum direct voltage corresponding to the peak value of the alternating voltage.
- When higher voltages are required voltage multiplier circuits are used. The common circuits are the voltage double circuit
- Used for higher voltages.
- Generate very high dc voltage from single supply transformer by extending the simple voltage doubler circuit.

ELECTRIC PROPULSION AND HIGH VOLTAGE PRACTICE

Marine Electric Propulsion

Integrated electric propulsion (IEP) or full electric propulsion (FEP) or integrated full electric propulsion (IFEP) is an arrangement of marine propulsion systems such that gas turbines or diesel generators or both generate three phase electricity which is then used to power electric motors turning either propellers.

It is a modification of the combined diesel-electric and gas propulsion system for ships which eliminates the need for clutches and reduces or eliminates the need for gearboxes by using electrical transmission rather than mechanical transmission of energy. Electric propulsion for many new ships is now re-established as the popular choice where the motor thrust is governed by electronic switching under computer control. The high power required for electric propulsion usually demands a high voltage (HV) power plant with its associated safety and testing procedures. Passenger ships have always been the largest commercial vessels with electric propulsion and, by their nature, the most glamorous. This should not, however, obscure the fact that a very wide variety of vessels have been, and are, built with electric propulsion.

Early large passenger vessels employed the turboelectric system which involves the use of variable speed, and therefore variable frequency, turbo-generator sets for the supply of electric power to the propulsion motors directly coupled to the propeller shafts. Hence, the generator/motor system was acting as a speed reducing transmission system. Electric power for auxiliary ship services required the use of separate constant frequency generator sets.

A system that has generating sets which can be used to provide power to both the propulsion system and ship services has obvious advantages, but this would have to be a fixed voltage and frequency system to satisfy the requirements of the ship service loads. The provision of high power variable speed drives from a fixed voltage and frequency supply has always presented problems. Also, when the required propulsion power was beyond the capacity of a single d.c. motor there was the complication of multiple motors per shaft. Developments in high power static converter equipment have presented a very convenient means of providing variable speed a.c. and d.c. drives at the largest ratings likely to be required in a/ marine propulsion system.

The electric propulsion of ships requires electric motors to drive the propellers and generator sets to supply the electric power. It may seem rather illogical to use electric generators, switchgear and motors between the prime-movers (e.g. diesel engines) and propeller when a gearbox or length of shaft could be all that is required.

There are obviously sound reasons why, for some installations, it is possible to justify the complication of electric propulsion:

1. Flexibility of layout
2. Load diversity between ship service load and propulsion
3. Economical part-load running
4. Ease of control
5. Low noise and vibration characteristics

FLEXIBILITY OF LAYOUT

The advantage of an electric transmission is that the prime-movers, and their generators, are not constrained to have any particular relationship with the load as a cable run is a very versatile transmission medium. In a ship propulsion system it is possible to mount the diesel engines, gas turbines etc., in locations best suited for them and their associated services, so they can be remote from the propeller shaft. Diesel generator sets in containers located on the vessel main deck have been used to provide propulsion power and some other vessels have had a 10 MW generator for ship propulsion duty mounted in a block at the stern of the vessel above the ro-ro deck.

Another example of the flexibility provided by an electric propulsion system is in a semi-submersible, with the generators on the main deck and the propulsion motors in the pontoons at the bottom of the support legs.

LOAD DIVERSITY

Certain types of vessels have a requirement for substantial amounts of electric power for ship services when the demands of the propulsion system are low. Tankers are one instance of this situation and any vessel with a substantial cargo discharging load also qualifies. Passenger vessels have a substantial electrical load which, although relatively constant, does involve a significant size of generator plant. There are advantages in having a single central power generation facility which can service the propulsion and all other ship loads as required.

ECONOMICAL PART-LOAD RUNNING

Again this is a concept that is best achieved when there is a central power generation system feeding propulsion and ship services, with passenger vessels being a good example.

It is likely that a typical installation would have between 4-8 diesel generator sets and with parallel operation of all the sets it becomes very easy to match the available generating capacity to the load demand. In a four engine installation for example, increasing the number of sets in operation from two that are fully loaded to three partially loaded will result in the three sets operating at a 67% load factor which is not ideal but also not a serious operating condition, It is not necessary to operate generating sets at part-load to provide the spare capacity to be able to cater for the sudden loss of a set, because propulsion load reduction may be available instantaneously, and in most vessels a short time reduction in propulsion power does not constitute a hazard.

The propulsion regulator will continuously monitor the present generator capability and any generator overload will immediately result in controlled power limitation to the propulsion motors. During manoeuvring, propulsion power requirements are below system capacity and failure of one generator is not likely to present a hazardous situation.

EASE OF CONTROL

The widespread use of controllable pitch propellers (cpp) has meant that the control facilities that were so readily available with electric drives are no longer able to command the same premium. Electric drives are capable of the most exacting demands with regard to dynamic performance which, in general, exceed by a very wide margin anything that is required of a ship propulsion system.

LOW NOISE

An electric motor is able to provide a drive with very low vibration characteristics and this is of importance in warships, oceanographic survey vessels and cruise ships where,/-for different reasons, a low noise signature is required. With warships and survey vessels it is noise into the water which is the critical factor whilst with cruise ships it is structure borne noise and vibration to the passenger spaces that has to be minimised.

For very high power, the most favoured option is to use a pair of high efficiency, high voltage a.c. synchronous motors with fixed pitch propellers (FPP) driven at variable speed by frequency control from electronic converters. A few installations have the combination of controllable pitch propellers (CPP) and a variable speed motor. Low/medium power propulsion(1-5 MW) may be delivered by a.c. induction motors with variable frequency converters or by d.c. motors with variable voltage converters. The prime-movers are conventionally constant speed diesel engines driving a.c. generators to give a fixed output frequency. Gas turbine driven prime- movers for the generators are likely to challenge the diesel option in the future. Conventionally, the propeller drive shaft is directly driven from the propulsion electric motor (PEM) from inside the ship. From experience obtained from smallerexternal drives, notably from ice-breakers, some very large propulsion motors are being fitted within rotating pods mounted outside of the ship's hull. These are generally referred to asazipods, as the whole pod unit can be rotated through 360° to apply the thrust in any horizontal direction, i.e. in azimuth. This means that a conventional steering plate and stern side-thrusters are not required.

Ship manoeuvrability is significantly enhanced by using azipods and the external propulsion unit releases some internal space for more cargo/passengers while further reducing hull vibration. Gradual progress in the science and application of superconductivity suggests that future generators and motors could be super-cooled to extremely low temperatures to cause electrical resistance to become zero.

Marine Electric Propulsion

- Podded drives offer greater propulsion efficiency and increased space within the hull by moving the propulsion motor outside the ships hull and placing it in a pod suspended underneath the hull.
- Podded drives are also capable of azimuth improving ship maneuverability. Indeed, podded drives have been widely adopted by the cruise ship community for these reasons.
- The motors being manufactured now are as large as 19.5 MW, and could provide the total propulsion power.
- In an AC drive, a frequency converter is used to control the speed and torque of electric motor. The speed of the AC electric motor can be controlled by varying the voltage and frequency of its supply. A frequency converter works by changing the constant frequency main electrical supply into a variable frequency output.
- The ideal simplicity of the induction motor, its perfect reversibility and other unique qualities render it eminently suitable for ship Propulsion.

ELECTRIC PROPULSION

- Diesel-Generator sets to produce electricity to common grid for propulsion and ship use.
- Variable speed drives to rotate fixed pitch propellers.
- Commonly used in Cruise vessels, LNG tankers, Off-shore vessels and Ice breaking vessels due to reduced fuel oil consumption, lower emissions and increased pay-load

• Large Diesel Engine for Main Engine
• Configuration of Electric Propulsion

- System for Ships
- Configuration of Electric Propulsion
- System for Ships
- Configuration of Electric Propulsion
- System for Ships
- Configuration of Electric Propulsion
- Configuration of Electric Propulsion
- System for Ships
- Configuration of Electric Propulsion
- System for Ships
- Configuration of Electric Propulsion
- System for Ships
- Comparison with Conventional and
- Electric Propulsion system

SHORE SUPPLY FACILITIES

Shore power supply facilities have adopted high voltage rather than low voltage by necessity in order to keep the physical size of related electrical equipment such as shore connection cables manageable.

Inevitably high voltage would otherwise introduce new risks to ship's crew and the shipboard installations if necessary safety features were not built into the HVSC system or safe operating procedures were not put in place.

Those onboard systems that are designed to accept high voltage shore power, typically involving the following things:

- Incoming power receptacles,
- Shore connection switchgear,
- Step-down transformer or isolation transformer,
- Fixed power cables,
- Incoming switchgear at the main switchboard and
- Associated instrumentation. HVSC is often referred to as cold ironing.

The system nominal voltage is considered to be in the range from 1 kv ac to 15 kv ac.

INFRASTRUCTURE CONSIDERATIONS

Electrical System Grounding Philosophy:

The manner in which electrical system is grounded (e.g., ungrounded system, solid neutral grounding system, low impedance neutral grounding system, or high impedance neutral grounding system), including ground potential transformer method. Circuit. protection strategy is built around the selected method of system grounding in terms of over voltage prevention, over current prevention or continued operability under single phase grounded condition.

SYSTEM GROUNDING COMPATIBILITY

Arrangements are to be provided so that when the shore connection is established, the resulting system grounding onboard is to be compatible with the vessel's original electrical system grounding philosophy (for instance, the shipboard ungrounded power distribution system is to remain ungrounded, or the shipboard high impedance grounding system is to remain high impedance grounded within the design grounding impedance values). Ground fault detection and protection is to remain available after the shore connection has been established.

Cable Management System:

The cable management system is the ship's interface point with the shore power system. The cable management system is typically composed of flexible hv cables with the plug that extends to the shore power receptacle, cable reel, automatic tension control system with associated control gears, and instrumentation. shore power is fed to the shore connection switchboard via the cable management system.

SHORE CONNECTION SWITCHBOARD

where no cable management system is provided onboard, the shore connection switchboard is normally the ship's interface point with the shore power system. hv shore power is connected to this shore connection switchboard by means of an hv plug and socket arrangement. the shore connection switchboard is provided with a shore power connecting circuit breaker with circuit protection devices.

ONBOARD RECEIVING SWITCHBOARD

The receiving switchboard is normally a part of the ship's main switchboard to which the shore power is fed from the shore connection switchboard.

EQUIPOTENTIAL BONDING

Equipotential bonding between the ship and the shore is to be provided. An interlock is provided such that the HV shore connection cannot be established until the equipotential bonding has been established. The bonding cable may be integrated into the HV shore power cable. If the equipotential bonding cable is intended to carry the shipboard ground fault current, the cable size is to be sufficient to carry the design maximum ground fault current.

LOAD TRANSFER

Temporary Parallel Running

Where the shipboard generator is intended to run in parallel with the shore power for a short period of time for the purpose of connecting to the shore power or back to ship power without going through a blackout period, the following requirements are to be complied with:

i) Means are to be provided to verify that the incoming voltage is within the range for which the shipboard generator can be adjusted with its automatic voltage regulator (AVR)
ii) Means are to be provided for automatic synchronization
iii) Load transfer is to be automatic
iv) The duration of the temporary parallel running is to be as short a period as practicable allowing for the safe transfer of the load. In determining the rate of the gradual load transfer, due regard is to be paid to the governor characteristics of shipboard generator in order not to cause excessive voltage drop and frequency dip.

Load Transfer Via Blackout

Where load transfer is executed via blackout (i.e., without temporary generator parallel running), safety interlock arrangements are to be provided so that the circuit breaker for the shore power at the shore connection switchboard cannot be closed while the HV switchboard is live with running shipboard generator(s).

Safety Interlocks

An interlock, which prevents plugging and unplugging of the HV plug and socket outlet arrangements while they are energized, is to be provided.

Handling of HV Plug

While the HV shore connection circuit breakers are in the open position, the conductors of the HV supply cables are to be automatically kept earthed by means of an earthing switch. A set of pilot contactors embedded in the HV plug and socket-outlet may be used for this purpose. The earthing switch control is to be designed based on a fail-to-safe concept such that the failure of the control system will not result in the closure of the earthing switch onto the live HV lines.

HV Shore Connection Circuit Breakers

Arrangements are to be provided to prevent the closing of the shore connection circuit breaker when any of the following conditions exist:

i) Equipotential bonding is not established
ii) The pilot contact circuit is not established
iii) Emergency shutdown facilities are activated
iv) An error within the HV connection system that could pose an unacceptable risk to the safe supply of shoreside power to the vessel. These errors may occur within the alarm system, whether on board

the ship or at the shoreside control position, or within any relevant safety systems including those which monitor system performance.

v) The HV supply is not present

HVSC Circuit Breaker Control:

HV shore connection circuit breakers are to be remotely operated away from the HVSC equipment. HV shore connection circuit breakers are to be made only when it has been established that personnel are evacuated from the HV shore connection equipment compartments. The operation manual is to describe these established procedures.

HVSC Emergency Shutdown:

In the event of an emergency, the HV system shall be provided with shutdown facilities that immediately open the shore connection circuit breaker. These emergency shutdown systems are to be automatically activated.

Any of the following conditions are to cause emergency shutdown of the shore power supply:

i) Loss of equipotential bonding
ii) High tension level of HV flexible shore connection cable, or low remaining cable length of cable management system
iii) Shore connection safety circuits fail
iv) The emergency stop button is used
v) Any attempts to disengage the HV plug while live (this may be achieved by the pilot contactors embedded in the plug and socket such that the pilot contactors disengage before the phase contactors can disengage)

TESTS OF HV SWITCHBOARDS

Type Test

HV switchboards are to be subjected to an AC withstand voltage test or other relevant national or international standards. A test is to be carried out at the manufacturer's test facility in the presence of the Surveyor.

Onboard Test

After installation onboard, the HV switchboard is to be subjected to an insulation resistance test in accordance with Table-2 in the presence of the Surveyor.

EQUIPMENT DESIGN:

Air Clearance

Phase-to-phase air clearances and phase-to-earth air clearances between non-insulated parts are to be not less than the minimum, as specified in Table

Creepage Distance

Creepage distances between live parts and between live parts and earthed metal parts are to be adequate for the nominal voltage of the system, due regard being paid to the comparative tracking index of insulating materials under moist conditions according to the IEC Publication 60112 and to the transient overvoltage developed by switching and fault conditions.

SHORE CONNECTION SWITCHBOARD:

Construction

The HV shore connection switchboard is to be designed, manufactured and tested in accordance with a recognized standard code of practice as given by IEC.

Circuit Breaker

i) Shore connection HV circuit breaker is to be equipped with low voltage protection (LVP)
ii) The rated short-circuit making capacity of the circuit breaker is not to be less than the prospective peak value of the short-circuit current
iii) The rated short-circuit breaking capacity of the circuit breaker is not to be less than the maximum prospective symmetrical short-circuit current
iv) HV shore connection circuit breaker is to be remotely operated
   a. HV Circuit Breakers may beAir-Break (scarcely used)
   b. Oil-Break (not used in ships)
   c. Gas-Break (SF – 6 – Sulphur Hexafluoride)
   d. Vacuum-Break (Most Popular)

Bibliography

Akber Ayub: *Marine Diesel Engines*, Ane Books, Delhi, 2015.

Alok Tripathi: *Maritime Heritage of Indian Ocean*, Sharada Publishing House, Delhi, 2013.

Ananta Narayan Mishra: *Glimpses of Maritime History of Ancient Odisha*, Kunal Books, Delhi, 2010.

Bell: *Diesel Engineering : Electricity and Electronics*, Cengage, 2009.

Bennett: *Diesel Engineering : Brakes, Steering and Suspension Systems*, Cengage, 2009.

Bennett: *Diesel Engineering : Electronic Diesel Engine Diagnosis*, Cengage, 2009.

Dixon: *Diesel Engineering : Heating, Ventilation, Air Conditioning and Refrigeration*, Cengage, 2009.

Gadepalli Ravi Kiran Sastry: *Bio-Diesel : Biodegradable Alternative Fuel for Diesel Engines*, Readworthy, 2008.

Gert Jan Bom, I H Rehman, David van Raalten, Rajeshwar Mishra and Frank van Steenbergen: *Technology Innovation and Promotion in Practice: Pumps, Channels, and Wells. Reducing Fuel Consumption, Emissions, and Costs*, TERI, 2002.

Gurpreet S Khurana: *Maritime Forces in Pursuit of National Security : Policy Imperatives for India*, Shipra Books, Delhi, 2008.

K. R. Singh: *Coastal Security : Maritime Dimensions of Indias Homeland*, Vij Books, Delhi, 2012.

K. Sridharan: *India's Maritime Heritage*, CBS Publication, Delhi, 2003.

K.M. Srinivasan: *Centrifugal Pumps*, New Age International, Delhi, 2016.

K.S. Pavithran: *Foreign Policy and Maritime Security of India*, New Century Publications, Delhi, 2013.

K.S.Mathew: *Imperial Rome, Indian Ocean Regions and Muziris: New Perspectives on Maritime Trade*, Manohar Publication, Delhi, 2015.

L.N. Swamy: *Maritime Contacts of Ancient India : With Special Reference to West Coast*, Harman Publication, Delhi, 2000.

Manoj Gupta: *Maritime Affairs : Integrated Management for India*, Manas Publication, Delhi, 2005.

MICHAEL: *Water Wells & Pumps, Second Edition*, Tata McGraw-Hill, 2008.

Pius Malekandathil: *Maritime India Trade, Religion and Polity in the Indian Ocean*, Primus Books, Delhi, 2010.

Pradeep Kaushiva and Kamlesh K. Agnihotri: *Indian and American Perspectives on Technological Developments in the Maritime Domain and Their Strategic Implications in the Indian Ocean Region*, KW Publishers, Delhi, 2013.

Ravi Vohra, P.K. Ghosh and D. Chakraborty: *Contemporary Transnational Challenges: International Maritime Connectivities (Indian Ocean Naval Symposium)*, KW Publication, Delhi, 2008.

Sila Tripati, A.S. Gaur and Sundaresh: *Maritime Archaeology and Shipwrecks off Goa*, Kaveri Books, Delhi, 2014.

Sila Tripati: *Maritime Archaeology for Beginners*, Kaveri Books, Delhi, 2009.

Sila Tripati: *Maritime Contacts of the Past: Deciphering Connections Amongst Communities*, Kaveri Books, Delhi, 2015.

Sugandha: *Evolution of Maritime Strategy and National Security of India*, Decent Books, Delhi, 2008.

V.S. Lobanoff and R. Ross: *Centrifugal Pumps : Designs and Application*, Jaico Publishing House, Delhi, 2005.

V.S. Lobanoff and R. Ross: *Centrifugal Pumps: Designs and Application*, Jaico Publishing House, Delhi, 2003.

Vijay Sakhuja: *Asian Maritime Power in the 21st Century : Strategic Transactions: China India and Southeast Asia*, Pentagon Press, New York, 2011.

Yogendra Kumar: *Diplomatic Dimension of Maritime Challenges for India in the 21st Century*, Pentagon Press, New York, 2015.

Index